# THE CHANGING GEOGRAPHY OF ASIA

*Edited by*
*Graham P. Chapman and*
*Kathleen M. Baker*

FOR THE DEPARTMENT OF GEOGRAPHY AT SOAS

London and New York

First published 1992
by Routledge
11 New Fetter Lane, London EC4P 4EE

Simultaneously published in the USA and Canada
by Routledge
a division of Routledge, Chapman and Hall, Inc.
29 West 35th Street, New York, NY 10001

© 1992 Graham P. Chapman and Kathleen M. Baker
Typeset in Linotron Garamond by
J&L Composition Ltd, Filey, North Yorkshire
Printed and bound in Great Britain by
Biddles Ltd, Guildford and King's Lynn

*British Library Cataloguing in Publication Data*
A catalogue record for this book is available from the British
Library.
ISBN 0–415–05707–8 (HB)
ISBN 0–415–05708–6 (PB)

*Library of Congress Cataloging in Publication Data*

The Changing Geography of Asia / edited by
Graham P. Chapman and Kathleen M. Baker.
Includes bibliographical references and index.
ISBN 0–415–05707–8 (HB). – ISBN 0–415–05708–6 (PB)
1. Asia–Geography.   I. Chapman, Graham. P.
II. Baker, Kathleen M., 1950–.
III. University of London, School of Oriental and African
Studies. Dept. of Geography.
DS5.92.C45      1992
915–dc20
91–44801   CIP

# CONTENTS

v

CONTENTS

# FIGURES

# TABLES

ix

# CONTRIBUTORS

Dr Kathleen Baker is Lecturer in Geography at the School of Oriental and African Studies

Dr Robert W. Bradnock is Lecturer in Geography at the School of Oriental and African Studies

Graham Chapman is Professor of Geography at the School of Oriental and African Studies

Dr Tamara Dragadze at the time of writing was ESRC Research Fellow at the School of Oriental and African Studies

Dr Richard Louis Edmonds is Lecturer in Geography at the School of Oriental and African Studies

Dr Michael Freeberne is Lecturer in Geography at the School of Oriental and African Studies

Dr Edwina Palmer, who took her PhD at the Department of Geography, School of Oriental and African Studies, is now Lecturer in Geography at the University of Canterbury, New Zealand.

Dr Jonathan Rigg is Lecturer in Geography at the School of Oriental and African Studies

Philip Stott is Senior Lecturer in Geography at the School of Oriental and African Studies

# DISCLAIMER FROM THE AUTHORS AND THE PUBLISHERS

Whilst we have made every effort to ensure that all maps, diagrams, figures and data are as accurate as possible, we do not claim to represent the official position of any government or other agency with regard to the position of any borders, to the names of any places, nor to statistics.

# PREFACE

The undergraduate programme in the Department of Geography at the School of Oriental and African Studies started in 1965, just over twenty-five years ago as we write. The department decided to celebrate this quarter of a century by writing a Geography of Change, on the areas of the world to which our study and teaching is committed. This comprehensive review is being published in this and the companion volume on *The Changing Geography of Africa and the Middle East*.

It has of course been a daunting exercise, particularly to compress into books of this length some sense of the detailed knowledge of their areas which each expert in the department embraces. We have tried to do so at a level which is accessible to undergraduates and to serious A-level students, and we believe that we have succeeded in our aim – although obviously it is for the reader to decide.

Higher education in the UK is passing through a period of great stress. Resources for research and publication are being continually eroded, and it is notable that in the UK as a whole the study of geography at school and university levels has retreated from the broader overseas perspective it used to have, sometimes into esoteric theory building, sometimes into empirical studies which stress the UK or other developed nations. Clearly though, the world is becoming more and more interdependent, no matter whether in trade or defence, finance or in recognition of global climatic change. We need now, more than ever, a well-educated and well-informed citizenry, whose understanding of the wider world we live in can inform the conduct of their lives, and who can contribute to public opinion formation on vital contemporary issues as well.

The knowledge that such opinions are founded on cannot be

fixed. The world is changing so rapidly, that new knowledge generation at higher rates is even more important now than in the past. This means that geography in the UK must revitalize its traditional world-wide vision, and must be given the resources to complete its mission. We hope that this book will stimulate some awareness of the magnitude and speed of change, and the seriousness of the problems of understanding now and in the future. Perhaps, with luck, we will stimulate some students to join the woefully small band of experts who have committed their careers to this end.

# ACKNOWLEDGEMENT

The authors would like to thank Mrs Catherine Lawrence for the good-natured way in which she produced such high-quality maps, often at short notice.

# 1

# INTRODUCTION
## Asia's future as seen in the mid-1960s
### *Graham Chapman*

In this the second volume of our two-volume review of the changing geography of Africa and Asia over the last twenty-five years, we are considering all those parts of Asia outside of the Middle East (which was included in the first volume). Until the early years after the Second World War most of this vast area was either under colonial rule or, like China, heavily in the thrall of outside powers. Although many commentators would have stated then that Soviet Asia was also in effect several colonial territories of a socialist but imperialist state, the commonest position was to accept the incorporation of these territories as constituent parts of the Union.

One of the two main Asian countries to have evaded colonialism, Japan, precipitated the war fought against the colonial powers, in their colonial territories. The turmoil of the bitter conflict was an important agent in accelerating the pressures for independence, and much of the area was self-governing by the 1950s or even earlier. This means that by our approximate starting date of 1965, these states had established governmental structures and policies designed to initiate or accelerate development of both of their peoples and of their territories. That these two 'developments' are not necessarily synonymous was clearly evident from the style in which the USSR sought to develop the resources of Siberia, but showed less regard for the man-power that by coercion as well as inducement was assigned to the task.

By then, Japan was clearly re-established as a modern industrial power, though it had not yet frightened the rest of the developed world with its advanced technology and massive trade surpluses. But equally, there were massive areas of poverty and backwardness which seemed hardly to have advanced at all. These areas included

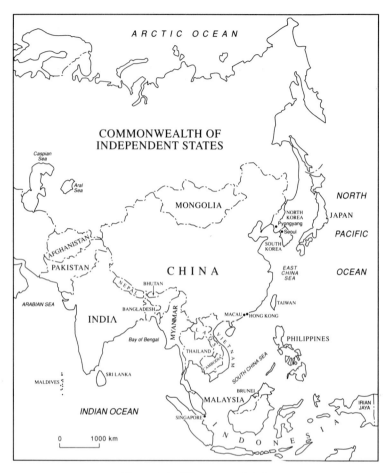

*Figure 1.1* The states of Asia

much of the Indian sub-continent, and much of China, though at
the time China was so closed that it was difficult to sort out fact
from official propaganda. The 'failure' of much of Asia to develop
swiftly was the subject matter of what was to rank as one of the
most massive and influential social science books written about Asia
in the post-war era – Gunnar Myrdal's *Asian Drama* (1968). The
book pulled few punches, and laid out quite clearly not only what
the author saw as the problems of these states, but also the problems
of the social scientists in commentating on these societies. The
failures were thus twofold, both real, on the ground, and in the

inability, as Myrdal saw it, for analysis to be either relevant or objective.

Writing ₁n the West on these regions had suffered from the colonial legacy, which had stressed analyses made for policy reasons by colonial government, and which had paid inadequate attention to the structure and functioning of the indigenous societies themselves. Where there had been such studies, they tended to be anthropological in approach and to describe a society which was assumed to be rather static. Classic Geographies, such as the Spate (1954) and Spate and Learmonth (1967) volume on India, Pakistan and Ceylon were also static in their approach – but in saying that, we are not denying their great value, much of which endures. Those authors who did attempt to write about development and change, Myrdal saw as imprisoned in the politics of the period. This had one major premise – that of the Cold War between the two superpowers, and their struggle to enlist client states in Asia. Because of this struggle, external writers colluded with the new governments, in twisting words. The countries which Myrdal persisted in calling 'underdeveloped' had instead become known as 'developing' countries, although he saw that term as applying better to the more rapidly changing economies of North America, Europe, and by then Japan as well.

> Impelled by the immense interests at stake, it is natural that the national authorities, the institutions sponsoring and financing research, and indeed public opinion in the West all press for studies of the problems of the underdeveloped countries. ... But the studies are also expected to reach opportune conclusions and to appear in a form which is regarded as advantageous, or at least not disadvantageous, to official and popular national interests. ... This same concern with national interests is also being seen in the underdeveloped countries themselves. Their institutions and their educated classes are becoming more and more touchy about questions dealing with social study.
>
> (Myrdal 1972: 5).

The effects of such an approach could be seen in the

> opportunistic arrangement of facts ... the use even in scholarly writings of labels like 'the free world' or 'the free Asian countries' to denote, not that people are free in the ordinary

sense of the word, but the purely negative fact that a country's foreign policy is not aligned to that of the communist bloc.

(Myrdal 1972: 6)

Economists in particular came in for criticism. Myrdal observed that they more than any other social scientists have been 'disposed to arrive at general propositions and then postulate them as valid for every time, place, and culture' (Myrdal 1972: 8). Restricted to the Western world, there might be value in such an approach. But for South Asia (by which Myrdal meant what we in this book call South Asia and South-East Asia) the general propositions 'do not fit'. And besides the use of inappropriate propositions, there was the equally pragmatic problem of a lack of reliable empirical data even when some categories of observation and labelling might have been nearly appropriate.

What Myrdal suggested was that we need research into the institutional and social structures which inhibited or permitted development, which were very different from those in the West. He could understand the reluctance of external writers to become involved in these issues, since it might seem close to a resurgence of the colonial writing which had justified the superiority of the colonizers by the inferior social customs of the colonized. But the pursuit of truth should not be obscured by discomfort, nor justified by the need for diplomacy. This he saw as valid in its own place, but if dominant in social science, then it was a disguised form of condescension.

It is of course difficult to talk of 'the state of Asia' in the mid-1960s having just heard Myrdal's view that objectivity is so hard to achieve, and that data are often missing or unreliable. But some commentators did try and achieve a broad-brush assessment, one of the most wide-ranging, yet brief and stimulating, being that by Bairoch (1975), who concentrated on change in the Third World in the period 1900–1960/70. His first chapter was on population growth. In it he outlined the dimensions of the truly explosive nature of that growth, and estimated it would peak at around 2.9 per cent. He saw little hope that family planning programmes would reduce the rate, since the population pyramid was so dominated by the younger age groups, yet to reach reproductive age. Although he thought that large populations could be fed, a major concern was that the rate of increase in food production would not keep abreast of the rate of population growth. The new technology of the Green

Revolution was known, but he commented that what mattered most was the rate at which institutional factors could adapt to the technology and diffuse it. He also noted that productivity levels in agriculture were much below those that had been achieved in Europe prior to its industrialization and urbanization. There the success of the preceding agricultural revolution had provided capital and spare man-power. He also noted that Europe's industries had the world as an export market.

In the mid-1960s then, a prime concern for Asia, as still for Africa now, was food security. Bairoch's approach to this was to set a standard of calorie output per agricultural worker which would enable the society to survive with a minimal risk of famines. His calorie level worked out at 4.9 million calories per worker per year. As Table 1.1 shows, much of Asia (and much of Africa) was just below or just above this figure. The parlous nature of the food supply was in fact confirmed, when India became desperately reliant on imports during the mid-1960s, and something near famine afflicted several parts of the country, and as China too underwent a famine whose dimensions are only now being truly revealed. It is no wonder that so much hope was being put by both international and national agencies on to the 'miracle' of the much-vaunted Green Revolution.

Other phenomena which Bairoch (1975: 149) highlighted included the discrepancy between the level of urbanization achieved in Asia (excluding Asiatic USSR), still low but increasing fast, and the level of manufacturing activity. Averaging out the figures for Europe (excluding the UK) for two years 1850 and 1880, we can take, even if a little crudely, Europe to have been about 13.5 per cent urban about the year 1865. At that date the average for the percentage employment in industry was about 17 per cent – therefore giving a ratio between the two figures of 0.8:1. By the 1930s, Europe, still excluding the UK, was 32 per cent urban and had 22 per cent of its active population engaged in industry – giving a ratio of 1.45:1. By comparison for Asia (excluding Japan) we quote just the figures for 1960 – which show an urbanization level of 13.7 per cent but only 9.0 per cent engaged in industry. The result is a ratio of 1.52:1. The implication in these figures is that urbanization in Asia is not following the European precedent, that somehow it is preceding an equivalent growth in the industrial base. It gives some credence to the impression of cities with 'prematurely' high levels of service activity or unrecorded petty manufacturing.

*Table 1.1* Productivity of agricultural workers (expressed in millions calories output per worker per year): averages 1960/64–1968/72 (unless date stated otherwise)

|  |  |  |
|---|---|---|
| Africa |  |  |
| Ghana | 3.9 |  |
| Kenya | 5.3 |  |
| Madagascar | 10.6 |  |
| Morocco | 6.3 |  |
| Nigeria | 3.6 |  |
| Tunisia | 3.6 |  |
| Zaire | 5.5 |  |
| Asia |  |  |
| China | 5.4/6.4[a] |  |
| Ceylon[b] | 5.4 |  |
| India | 4.0 |  |
| Pakistan[c] | 4.3 |  |
| Indonesia | 4.5 |  |
| Philippines | 4.7 |  |
| Thailand | 7.0 |  |
| Comparisons |  |  |
| Britain in 1810 | 14.0 |  |
| USA in 1840 | 21.5 |  |
| USA in 1968–72 | 330.0 |  |

*Source:* Bairoch (1975: Tables 9 and 13)
*Notes:* [a] The first figure is based on Western estimates, the second on official government sources.
[b] Now Sri Lanka.
[c] Includes East Pakistan, now Bangladesh.

The earlier figure for Europe also suggests that much manufacturing activity preceded urbanization and was distributed in rural areas (although that might beg many definitional questions).

Manufacturing featured low in the exports of Asia as a whole. (Again this statement excludes Asiatic USSR. That region in fact 'exported' industrial raw materials, energy, part-processed materials and finished goods, overwhelmingly to European USSR.). The quality of goods could not compete with the established metropoles of the North, and in addition tariffs in the importing countries of the North discriminated more against the value-added nature of manufactured exports than against primary exports. It is not surprising that many countries in the 1960s saw their future more in terms of self-sufficient industrialization and import substitution than in opening themselves more widely to continued domination in a neo-colonialist subservient position. There were observers (e.g. Little et al. 1970) who doubted whether such import-substitution would enable industry to match world efficiency levels, and to prepare these countries for international competitiveness, but external criticism was not welcomed by governments, most of which had associated their authority with the cause of national independence in all spheres.

In the case of Asiatic USSR the key question as seen from Moscow was how to exploit the resources of the regions. The question as to whether it was economically sensible to do this seemed less important than working out how to do it. This attitude is something that is ascribed in the first instance to communism's belief that humans should and will have dominion over nature, that political and human relationships have a material foundation. But it is wrong to think of this as an attitude that has its origins in communism. It was part of the Protestant ethic too, which religious settlers took to the USA, and which they used to justify their rights to alienate land from the Indians, who must forgo their claims to territory since they did not set about taming and subjugating the wilderness. Since the exploitation of these remote regions was expensive, it might have been an obvious consequence that little attention would be paid to environmental costs incurred. But public discussion of such problems was not allowed in the USSR, nor even in the West in the mid-1960s was there the same kind of alarm and concern that are now commonplace.

For most of Asia the key questions which the 1960s therefore posed were: first, food security; second, the rate of growth of the industrial economies and the relationship of this to rates of urbanization; third, the nature and style of institutional and cultural change; and fourth, linked to all the above points, the degree and manner of change in the inclusion of most of these states in the world economic system. Two regions were clearly going to 'go it alone' – and to that extent the inclusion of China and the USSR in the world trading system was not a major issue. Population growth rates also fascinated, or appalled, observers, but were on the whole taken to be a 'given' for many countries, with the possible exception of China. The thesis was simply that the rates would diminish in the style of a classic demographic transition, as a result of development occurring.

The remaining pages of this volume explore these and many other themes for the countries of Asia. That they have been written by experts all associated with the School of Oriental and African

Studies is worthy of note. Phillips's review of the status of Asian studies in universities in the UK sketched the interplay of academic study and political pressure in the post-war era. He commented that the Scarborough report of 1947 had observed that 'in peace as in war a knowledge of Asian countries had to be given a permanent and growing place in British culture' (Phillips 1967: 1). The base for broad and general studies did not then exist. The tradition of Orientalism, based around the study of classic Asian languages, did exist, along with a residue of political knowledge from the ex-administrators of Empire. But, 'little claim was made on the social sciences, and in any event, the social scientists themselves were still too pre-occupied by the struggle to introduce and consolidate Western social studies in British Universities to venture boldly on to the rather unfamiliar, difficult, and certainly expensive field of Asian Studies' (Phillips 1967: 3). But unless this capacity was somehow increased, it would not be possible to consider such questions as

> Who were the Indian Middle Classes? How had they arisen? ... Although we are on occasion told about the increase in commodity exports, we are rarely introduced to the equally important rise and fall in internal consumption. What exactly was happening to Indian society? What was the relation between caste and growing industry?
>
> (Phillips 1967: 3)

The Hayter Report of 1961 took a more specific and expansionist approach and out of the funding which the government subsequently made available was born at the School of Oriental and African Studies a new Department of Economic and Political Studies, a Department of Sociology and a Department of Geography. Special posts were also funded in other universities which had relevant infrastructural support. The Department of Geography at SOAS of which the authors of this volume are members, or with which they are associated, would not have existed but for this action, and many of us would either never have been attracted into these fields of studies, or have drifted away for lack of support. Phillips wrote (1967: 3) 'There is a striking absence of fully synthesised and balanced studies making adequate reference to economic and social factors, a situation which compares unfavourably with the plethora of good political writing on policy matters.' We hope that we are beginning to redress the 'lopsidedness' which Phillips detected.

Myrdal concluded his introduction by saying:

> In the classic conception of drama – as in the theoretical phase of a scientific study – the will of the actors was confined within the shackles of determinism. The outcome at the final curtain was predetermined by the opening up of the drama in the first act, accounting for all the conditions and causes of later developments. But in life while the drama is still unfolding the will is instead assumed to be free, within limits, to choose between alternative courses of action. History, then, is not to be taken to be predetermined. Rather it is within the power of man to shape it. And the drama thus conceived is not necessarily tragedy.
>
> (Myrdal 1972: 16)

We subscribe to Myrdal's view that independent social scientists should be working as objectively as possible on the development of Asia, but we recognize that the knowledge base in terms of personnel as well as data is very narrow. Within the confines of expertise that we have available, we hope we will have gone some way in this volume to considering the major changes, some dramatic, in the societies of Asia.

## REFERENCES

Bairoch, P. (1975) *The Economic Development of the Third World since 1900*, translated from the French by C. Postan, London: Methuen.

Little, I., Scitovsky, T. and Scott, M. (1970) *Industry and Trade in Some Developing Countries*, London: Oxford University Press.

Myrdal, G. (1968) *Asian Drama: An Enquiry into the Poverty of Nations*, (abridgement 1972 by S.S. King), London: Allen Lane.

Phillips, C.H. (1967) 'Modern Asian studies in the universities of the United Kingdom', *Modern Asian Studies* 1(1): 1–14.

Spate, O.H.K. (1954: 1st edn); Spate, O.H.K. and Learmonth, A.T.A. (1967: 3rd edn) *India and Pakistan: A General and Regional Geography*, London: Methuen.

# 2

# CHANGE IN THE SOUTH ASIAN CORE
## Patterns of growth and stagnation in India
### Graham Chapman

## INTRODUCTION

The name 'India' comes from the same root as 'Indus', the great river that flows out of the Himalayas in Kashmir, through Pakistan to the Arabian Sea. It is the name which non-Indians have applied to the peoples (H-indus) living by and beyond the Indus, and by transference also to the land. The word India has therefore always had a larger connotation than just the current Republic of India, and embraces most of the territory of South Asia that used to be embraced by the British Empire (see Figure 2.1). This greater India is well defined in terms of topography; it is the Indian sub-continent, hemmed in by the Himalayas on the north, the Hindu Khush in the west and the Arakanese in the east. Modern informa-tion on plate tectonics and continental drift has shown how these mountains have all been formed by the dent which the northward-moving Deccan block has made in the southern rim of Asia. It is still pressing in at the rate of about 6 cm a year, and the Himalayas are still being uplifted at about 6 cms a year. Thus, despite variations in climate from arid desert to tropical humid forest, this is a well-defined region – and it has been easily and simply defined as a geopolitical region as well, like Europe or North America. This means it is a region in which there are coherent bonds of culture and a continuous arena for trade and economic integration (but this is a permissive rather than deterministic statement). It is a natural arena within which to exert military power: i.e. its logical defensive perimeter is well defined. Yet despite all this, 'Greater India' (South Asia) is divided by religion. Pakistan and Bangladesh are over-whelmingly Muslim, and have Muslim constitutions. India is 80 per

10

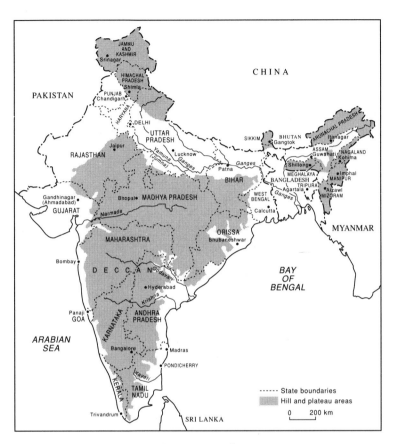

*Figure 2.1* India

cent Hindu, but has still managed to cling to a secular constitution. The borders between these states are in a geographical sense arbitrary – the product of a hasty attempt at independence in 1947 to separate the hostile communities of Hindus and Muslims. The lines wandered across rivers, roads and railways, tearing off what had been integrated territory, rather like East Germany was once summarily sealed off from West Germany.

The Republic of India is thus the largest of the sovereign states of a region which is recognized everywhere as the sub-continent of South Asia. It is the largest territorially, and also overwhelmingly the largest in population. Currently the size of its population is

11

second behind China: but many predict that some time in the next century it will be the first.

## POPULATION

In many parts of the Third World, data even on basic issues like population are either missing or too suspect to use with confidence. India has a remarkably good and extensive decennial census, which, although no doubt subject to error on many details, is nevertheless reliable enough to chart some of the broad demography. In 1961 India had a population of about 440 million, and a growth rate of about 2.0 per cent per annum. At that rate of growth it would have been easy to predict that by 1990 the population would be about 775 million, a figure which can be put into perspective by saying that the increment between the two dates of 335 million would have been the same as the total population of the USA and Mexico in 1990. All governments since Independence have seen this growth rate as a major threat, continually eroding the benefits of economic growth. They have all planned to reduce it by programmes to promote birth control, which in the Emergency period of 1975 even briefly reached the extreme of the enforced sterilization of males in some areas. But there has been no campaign of co-ordinated propaganda and financial discrimination such as attempted in China and credited there with a major effect on birth rates. In India it is now clear that the population growth rate actually accelerated in these thirty years, the actual total for 1990 being about 822 million (Muthiah 1987: 16). The birth rate has dropped from above thirty-seven per thousand to about thirty-two now, but the death rate has dropped faster, from more than sixteen per thousand to a little over ten. The population growth rate, having surged to 2.25 per cent in the late 1970s, is only now coming back down towards about 2.00 per cent. Thus there are at last signs that the demographic transition from a high birth rate plus high death rate regime to a low rate for both is occurring, but slowly.

The reason why change is so slow is that so many cultural norms have to change, much more than they have done in the recent past. Some sections of India's population, such as the 120 million Muslims, belong to creeds which like the Catholics disapprove of artificial birth control. Virtually the whole of society places a great value on male successors, and hence there is an imperative to continue having children if the first births to a married couple are girls. Above all it is recognized that early marriage of girls, which is still the norm for most and certainly prevalent in rural areas, and the higher level of illiteracy amongst women, contributes to their dependent and domestic role in life, and deprives many of either the knowledge or will to practise family planning. The average number of children per couple is still around five (some authors put it still at nearer six). These children can be both an asset, helping with work or earning small amounts of money, providing security for their parents' old age, and a liability, since girls for example require dowries in the majority of marriages. Children who are sick and need care are also a liability. The early marriage of a daughter can therefore be an attractive proposition for a poor family, thus perpetuating the cycle of greater fertility throughout a longer child-bearing marriage. It also seems likely that the perceived risk of children dying is still far too high, particularly in rural areas. The death rate for children under 5 in rural areas is still about 25 per cent. For those who survive these critical early years life expectancy is as high as it is in Britain. It is mainly infant deaths that produce the low average life expectancy.

For the population to stabilize the average number of children per couple would have to decline to around 2.3 from the current 5. It is true that there is a new urban middle class which has a lower birth rate, and perhaps gets nearer the government's ideal family of four (the slogan is 'We two, our two'), but this class is still only a small part of the total population. Marriage remains overwhelmingly by parental arrangement, even if there is a little more opportunity for the children to express their views of possible spouses, and it is still invariably within caste. The extended family and the security of living with many relatives that it brings is still very much the norm. All of this suggests that it is not easy for a couple to express independent views about the number of children they should conceive. Tradition is still a formidable obstacle to change. But there are signs that urbanization, seen by many as undesirable, does nevertheless have a beneficial effect on population growth. Urban birth and death rates are both significantly lower than rural ones: in 1987 the birth rates were 27.1 and 33.5 respectively, and the death rates 7.3 and 11.9 (Tata Services Ltd 1989: 38).

## URBANIZATION

In absolute terms, India's urban areas have shown an accelerating rate of growth over the last three decades – from a decadal increase of 30.2 million 1961–71, to 50.6 million in 1971–81, and then to an estimated 70.4 million increment in the nine years 1981–90. That is, a growth in new urban population of more than the entire population of the United Kingdom in just nine years. These people have to be housed, have to find some sort of living if not orthodox employment, and of course food. It is difficult to be precise how much of the increase has come from immigration from rural to urban areas, and how much from natural rates of increase in urban areas. On some calculations in the period 1971–81 a low estimate would be that there were 25.5 million rural–urban migrants in the total increment of 50.6 million, although some studies quote figures over 30 million. In broad terms about half the increase comes from each source, and this in turn means that in fact the rate at which India is urbanizing – expressed as an increase in the proportion of the total population resident in urban areas – is actually not very high. The urban percentage has gone from 18.2 per cent in 1961 to 23.7 in 1981 and is about 28 per cent now. In many Third World countries this figure has increased much faster – there is an ogive of expected growth which would lead one to expect perhaps a more rapidly accelerating trend. India thus remains overwhelmingly rural, despite the massive total urban population of 230 million today – almost the size of the US population.

Those who see the mass squalor of many urban areas, the congestion and pollution of the large cities, would wish that the process of urbanization stalled, or at the least concentrated in newer and smaller urban places. But there are very strong protagonists for the counter-argument, that indeed the current trends ought to be encouraged, because urban areas are seen to be more efficient at providing rising incomes, with better external economies and hence better local income multipliers for each unit of capital invested. The composition of the GNP has changed remarkably since independence, a trend that has accelerated through the last three decades. If one takes rural areas as synonymous with the agricultural part of the GNP, then the 80 per cent (1961) who were rural then and contributed just 50 per cent of the GNP, has now shifted to 72 per cent who provide only 29 per cent of GNP. But the other side of the coin is not easy to assess: industry has grown from around 20 per cent of GNP to only 29 or 30 per cent now. The gap has been made up by an increase in the value of services, some such as transport and communications essential to a growing economy. But many of these services are seen to be parasitic, a myriad of merchants, small

14

shopkeepers and hawkers, who are not efficient users of resources. But it is this middle sector, and the small-scale 'informal' sector, that is revealed only in local studies, not in national statistics. Detractors also point to the remarkable fact that in modern industry, despite all the investment, employment has stayed remarkably fixed at about 10 per cent. The argument then runs that modern technology, influenced if not dominated by Western expertise, is inherently unsuited to providing the right capital/labour mix for India. Urban areas are therefore bound to express a duality between industries under-using labour, and a complex of compensating activity which maximizes labour input per item of turnover, or per unit of input material (such as recycled waste).

## AGRICULTURE AND FOOD SUPPLY

In the mid-1960s, India suffered the worst droughts since independence in 1947. Famine threatened, and in some parts of the country it is probable that significant increases in mortality occurred. India became dependent on food imports, which reached a record of just over 10 per cent of food availability (Johnson 1979: 29) – in an overwhelmingly agricultural country. The problem was that agricultural output had been increasing less fast than population, and there was no margin to spare in the event of environmental hazards. Since then, the situation appears to have been radically transformed, at least at the aggregate level. The much-vaunted Green Revolution has taken off, and by the mid-1980s it appeared that India was massing food mountains. Public criticism, tinged with a certain pride at being in an EC-like predicament, of the cost of these mountains was rife, but drought and a downturn in production rapidly rectified the situation in 1987 and 1988. This time the vicissitudes were surmounted without recourse to bulk imports. This was achieved because from the late 1960s or early 1970s to 1990 the underlying increase in food grains output exceeded the population growth rate for the first time since independence. The index of agricultural production which stood at 100 in 1969 (averaged over 1967–70) had reached somewhere around 150 by 1987, indicating a slight increase in overall availability per capita (Tata Services Ltd 1989: 1). But the detailed picture is far less satisfactory. Wheat production had increased significantly, but rice production had not quite kept pace with population growth, and production in gram (pulses), the most important source of protein in vegetarian diets, and in oil-seeds for cooking, had completely stagnated, implying a deteriorating standard of diet. There is increasing concern that despite government programmes which aim to sell rationed quantities of food cheaply through Fair Price Shops, there is an increasing number of poor, particularly in rural areas, who simply do not get adequate access to food: their position in society does not allow them an 'entitlement'.

## ECONOMIC GROWTH AND STAGNATION IN TWO-SPEED INDIA

It makes little sense to discuss other changes at the all-India level, since regional differences are so strong. What is most significant is the way in which the regional patterns correlate over so many factors: lower rates of urbanization, poorer agriculture, slower growth in industry, poorer infrastructure, all combining in some backward areas, and the reverse in the richer areas. The inequalities have, if anything, increased. Nor is there a mechanism for reducing them, since the central government is relatively weaker now, and the local state governments stronger. Reallocation of resources between states is subject to a formula (known as the Gadgil Formula and bearing some methodological similarities to the calculation of Standardized Spending Assessments of local authorities in the UK), which while on the one hand making some adjustment for backwardness, also makes other adjustments in terms of local taxation success, which merely pushes the calculation back in the other direction. National planning still exists, but more so in terms of setting some broad strategic ideals, and the actual planning is now effectively devolved down to the states – and those which are backward are often less able to produce either the resources or the trained staff to do as well as the more progressive states.

The broad macro-regionalization which follows is based on state aggregates (see Tables 2.1, 2.2 and 2.3). As such it is not very satisfactory, since within states there are some great regional variations, but it does have the advantage of allowing the use of some grouped statistical data. Where there are very wide inequalities within states, the reader's attention will be alerted. The regionalization identifies a distinction between High Speed India (subdivided into three regions), and Low Speed India, dealt with here as a single unit. It just so happens that the population totals of these two macro-regions are about the same. So half of India is

*Table 2.1* Population and growth rates in the states and union territories of India

| | Population 1961 (mn) | Population 1990 (mn) | Per Capita GDP/% annual growth rate (1961–81) | Annual growth rate rural population (1961–81) | Annual growth rate urban population (1961–81) |
|---|---|---|---|---|---|
| *High speed* | | | | | |
| (a) North-west | | | | | |
| Haryana | 7.6 | 16.2 | 2.5 | 2.0 | 4.8 |
| Himachal Pradesh | 2.9 | 5.0 | | 2.1 | 3.0 |
| Jammu and Kashmir | 3.5 | 7.3 | 1.3 | 2.3 | 3.9 |
| Punjab | 11.2 | 19.6 | 3.0 | 1.6 | 3.8 |
| Delhi | 2.7 | 8.9 | | 0.8 | 4.7 |
| Chandigarh | 0.2 | 0.7 | | 1.7 | 6.1 |
| Rajasthan | 20.2 | 43.5 | 0.4 | 2.5 | 4.7 |
| (b) West | | | | | |
| Gujarat | 20.6 | 40.4 | 1.2 | 2.0 | 3.5 |
| Maharashtra | 39.5 | 74.2 | 1.6 | 1.6 | 3.4 |
| Dadra and Havar Nageli | | 0.2 | | 2.7 | |
| Goa, Daman Diu | 0.7 | 1.4 | | 1.4 | 4.7 |
| (c) South | | | | | |
| Andhra Pradesh | 36.0 | 63.2 | 1.2 | 1.6 | 4.0 |
| Karnataka | 23.6 | 44.6 | 2.2 | 1.8 | 4.2 |
| Kerala | 16.9 | 29.7 | 1.0 | 1.5 | 3.3 |
| Tamil Nadu | 33.7 | 55.7 | 0.9 | 1.2 | 2.5 |
| Pondicherry | 0.4 | 0.7 | | 0.5 | 4.8 |
| *Low speed* | | | | | |
| Bihar | 46.5 | 84.7 | −1.2 | 1.9 | 4.5 |
| Madhya Pradesh | 32.4 | 63.1 | 0.0 | 1.8 | 4.6 |
| Orissa | 17.5 | 30.9 | 1.0 | 1.5 | 5.4 |
| Uttar Pradesh | 73.7 | 133.7 | 0.4 | 1.8 | 4.9 |
| West Bengal | 34.9 | 64.8 | 0.0 | 1.9 | 2.8 |
| Sikkim | | 0.4 | | 3.4 | 10.0 |
| *Special North-east* | | | | | |
| Arunachal Pradesh | 0.4 | 0.8 | | 2.8 | 9.1 |
| Assam | 11.2 | 24.5 | | | |
| Manipur | 0.8 | 1.7 | | 1.2 | 10.3 |
| Meghalaya | 0.8 | 1.7 | | 2.4 | 5.1 |
| Mizoram | | 0.7 | | 2.4 | 12.4 |
| Nagaland | 0.4 | 1.1 | | 3.5 | 8.9 |
| Tripura | 1.2 | 2.5 | | 2.7 | 3.3 |
| India | 439.5 | 822.2 | | | |

*Sources:* Muthiah (1987); Tata Services Ltd (1989).
*Note:* Blanks indicate no data available.

*Table 2.2* Social and economic indicators in the states and union territories of India

| | Population below poverty line (1984–5) (%) | Value added in Rs per year per worker in agriculture (1981–2) | Value added in Rs per year per worker in industry (1981–2) | Value added in industry in Rs bn (1982–3) | Electric power generation Mw (1986) |
|---|---|---|---|---|---|
| *High speed* | | | | | |
| (a) North-west | | | | | |
| Haryana | 17 | 7,498 | 21,961 | 5,200 | 525 |
| Himachal Pradesh | 19 | 3,069 | 10,000 | 1,360 | 306 |
| Jammu and Kashmir | 23 | 3,888 | 37,917 | 360 | 197 |
| Punjab | 13 | 8,219 | 15,867 | 4,580 | 1,194 |
| Delhi | | | | | 1,016 |
| Chandigarh | | | | | |
| Rajasthan | 34 | 3,569 | 18,317 | 3,950 | 981 |
| (b) West | | | | | |
| Gujarat | 23 | 3,676 | 18,380 | 15,240 | 3,283 |
| Maharashtra | 34 | 2,826 | 25,382 | 36,010 | 7,201 |
| Dadra and Havar Nageli | | | | | |
| Goa, Daman | | | | | |
| Diu | | | | | |
| (c) South | | | | | |
| Andhra Pradesh | 35 | 2,683 | 9,380 | 10,160 | 3,987 |
| Karnataka | 33 | 2,881 | 18,365 | 8,140 | 2,502 |
| Kerala | 25 | 4,744 | 14,781 | 4,840 | 1,272 |
| Tamil Nadu | 37 | 1,597 | 16,883 | 16,910 | 3,819 |
| Pondicherry | | | | | |
| *Low speed* | | | | | |
| Bihar | 48 | 1,935 | 24,785 | 11,230 | 1,575 |
| Madhya Pradesh | 46 | 1,956 | 23,061 | 9,560 | 3,657 |
| Orissa | 42 | 3,137 | 16,500 | 3,610 | 1,200 |
| Uttar Pradesh | 43 | 2,642 | 19,038 | 14,950 | 5,375 |
| West Bengal | 48 | 3,225 | 15,505 | 16,340 | 2,758 |
| Sikkim | | | | | |
| *Special North-east* | | | | | |
| Arunachal Pradesh | | | | | |
| Assam | 38 | | 11,475 | 1,530 | 410 |
| Manipur | 23 | 2,180 | 5,000 | | 105 |
| Meghalaya | 37 | | | 90 | 175 |
| Mizoram | | | | | |
| Nagaland | | | | | 27 |
| Tripura | 48 | 2,877 | 2,857 | 70 | |
| India | 37 | 2,740 | 18,424 | 168,030 | |

*Sources:* Muthiah (1987); Tata Services Ltd (1989).
*Note:* Blanks indicate no data available.

*Table 2.3* Summary statistics for High Speed and Low Speed India

| | Population 1961 (mn) | Population 1971 (mn) | Population 1981 (mn) | Population 1990 (mn)[a] | Average annual growth rate per capita GDP/% 1961–81 | Population urban 1961 (%) | Population urban 1981 (%) | Population below poverty line 1984–5 | Villages electrified 1986 (%) | Villages with all-weather roads 1983 (%) |
|---|---|---|---|---|---|---|---|---|---|---|
| **High Speed** | | | | | | | | | | |
| (a) North-west | 48.3 | 61.9 | 80.9 | 101.2 | 1.5 | 22.5 | 27.9 | 22.5 | 79.4 | 46.8 |
| (b) West | 60.8 | 78.1 | 98.1 | 116.2 | 1.5 | 27.2 | 33.6 | 29.8 | 93.0 | 42.1 |
| (c) South | 110.6 | 135.8 | 165.1 | 193.9 | 1.3 | 20.9 | 26.8 | 33.5 | 81.0 | 41.7 |
| Low Speed | 205.0 | 252.6 | 314.2 | 377.6 | 0.0 | 13.6 | 18.1 | 45.4 | 54.1 | 19.4 |

*Source:* Calculated from Tables 2.1 and 2.2 and other data from Muthiah (1987); Tata Services Ltd (1989).
*Note:* [a] Estimate.

undoubtedly benefiting from good economic growth and rising real incomes: while the other half is stagnating, almost in a sea of despond. Clearly selective use of evidence can demonstrate that India is either a resounding success, or a catastrophic failure: unselective evidence shows that it is both.

The three sub-regions of High Speed India are first, in broad terms the north-west, i.e. Punjab, Haryana, Delhi Union Territory, north parts of Rajasthan, and western parts of Uttar Pradesh near Delhi, which form one 'success' region. For statistical purposes the whole of Rajasthan, which also embraces poor desert areas, is grouped with this region. For statistical purposes, Kashmir and Himachal Pradesh have also been included in this group. Another sub-region is the arc stretching from Bombay (Maharahstra) to Gujarat, encompassing many 'successful' urban regions. For statistical purposes the whole of Maharashtra is included – although the plateau areas of the Deccan are in parts backward and poor. The third sub-region is the group of southern Karnataka, much of Tamil Nadu, and parts of Kerala and the coastal region of the Krishna-Godavari delta in Andhra Pradesh. Again Andhra Pradesh includes Deccan areas which are much poorer than the capital region around Hyderabad or the coastal zone.

Low Speed India includes much of the central Deccan, less densely settled, and aggregated into the statistics here by placing Madhya Pradesh and Orissa in the region, together with the contrastingly very densely settled areas of eastern Uttar Pradesh, Bihar and West Bengal. For statistical purposes the whole of Uttar Pradesh is included.

This regional breakdown omits some areas – the north-eastern states of Assam, Nagaland, Manipur, Mizoram, Meghalaya and Arunachal Pradesh. These are in many senses problematical: apart from Assam (24 million) their populations are small. They are all areas of political turmoil and strong secessionist movements, but equally they are all boundary areas of strategic importance. Interestingly they are areas which have seen some of the most rapid urbanization in India in the 1980s. But all these factors are related, as the central government provides extra expenditure to retain local commitment, and as units of the army are deployed, and as rural areas become zones of insurgency conflict. These 'special' areas are not included in the general high-speed/low-speed dichotomization.

# HIGH SPEED INDIA

## The north-west

This is the oft-quoted 'miracle' food-basket of India. Here in Punjab and Haryana the new high yielding varieties of wheat launched the Green Revolution. From 1953 to 1986 wheat yields in Haryana grew by 330 per cent, in Punjab 196 per cent (Muthiah 1987: 158). These successes matched those of wheat growing areas in the developed world, and yields of 3,000 kg per hectare are respectable by international comparison. But there have been specific reasons why this should have been a success story. In Punjab in particular the average land holding per farm is twice that of Bihar, the development of irrigation which has been a continuing process since the 1850s, more advanced, and the commercial drive of the Sikh and Jat farming communities more pronounced than their equivalents elsewhere. This community has been helped by a sense of outward connection that is missing elsewhere – either with the concept of an 'official' infrastructure that has run the canal schemes, the agricultural research stations and the extension services, or with the emigrants from Punjab who have departed abroad, but whose continued connection with their homeland includes the very direct and significant benefit of remittances of capital. The links with the urban sector are much stronger than in most other areas. Not only are there proportionately more medium-sized small towns with their attendant regulated food grain markets, but also the rural road infrastructure and rural electrification, important for energizing the tube-wells that complement canal irrigation, are much better developed. This region is almost unique in India in that here the percentage of people living below the poverty line is less in the rural areas (10.9 per cent in Punjab and 15.2 per cent in Haryana) than in urban areas (21.0 and 16.9 per cent respectively – the only other cases are in Kerala and Manipur.) If there is a blot on the horizon it is that the yield increases promoted by the Green Revolution are levelling off, and that until there is another equivalent surge in productivity, there is always the threat that population growth will catch up with the gains made. There is another problem too: the new rice varieties, which should have helped Slow Speed India more, have turned out to be particularly well suited to the sunnier and well-irrigated areas of the north-west. From 1953 to 1986 rice yields in India as a whole increased by 83 per cent: but in Punjab by 289 per cent. However, rice is a 'wet-irrigated' as opposed to a 'dry-irrigated' crop – i.e. its water demands are very high. This means that the Punjab could face increasing tension over the availability and distribution of water, and in fact this tension is credited as being part of the problem in relation to the Punjabi separatist movement.

The region includes Delhi. This is the national capital region, which has accrued to it all the public employment and expenditure of a major capital city, and which could in the next century be India's largest city. In the two decades 1961–71 and 1971–81 it grew by 55 per cent and 57 per cent, from 2.4 million to 5.7 million people, and projections for the year 2001 place it alongside Bombay with a population of about 13 million. Nor is it just an administrative capital anymore, it is also becoming one of the country's main manufacturing centres, serving, it is true, to a large extent the consumer needs of the wealthier classes.

Not surprisingly, with such a huge increase in population, many people live in what are commonly called slums. These may include illegal squatter colonies, or land which is legally occupied but which has buildings of poor construction and virtually no services. For the whole of India the slum population is expressed as about 23 per cent of the total urban population. For Delhi, it is 47.5 per cent. Such slums are low-rise, yet have such high densities that they cover proportionately little of the built-up area, and they tend to be on peripheral sites.

To the middle classes of the more spacious 'colonies' (what in Britain would be termed 'estates') and to the bureaucracy, such slums are potentially threatening. 'If cities do not begin to deal constructively with poverty, poverty may well begin to deal more destructively with cities' (National Institute of Urban Affairs 1988: 65). But in a democracy they can be a political vote bank, and hence they do have bargaining power, sometimes asking for illegal squats to be legalized, sometimes to get better electric power supplies, sometimes to get water and sanitation. And surveys of slum dwellers record many positive responses to their lifestyles. Proportionately more of the slum dwellers than the rest of the population are rural–urban migrants. Many are from Uttar Pradesh in Delhi's case, and from poor and backward classes. Most find less discrimination, and an enhancement of their opportunities in life within the urban area. Many indeed accumulate small surpluses of income, and remit regularly a little money to their relatives still in the home villages. Studies are beginning to reveal just how many of the rural poor are dependent on urban income support in this way.

The rural areas have a well-developed infrastructure, comparatively speaking. The figure of 46.8 per cent of villages having all weather roads would be nearer 70 per cent if the rural areas of Rajasthan were excluded, and the figure for electrified villages also nearer the 90 per cent recorded for the west region, if the same area were excluded. This electrification should not be interpreted simply. It does not mean that all houses are connected – many could afford neither the connection nor the tariffs. And the supply in even the most advanced states is so erratic that any process which relies on mechanical power cannot rely on public electricity supplies only. Portable generators act as back-ups for the rich, and for many farmers a diesel engine is a desirable extra for energizing pumps.

The significance of roads is more easily interpreted. Few people can afford motorized vehicles: but wherever there are roads there are bus services, and shared taxis and motor-rickshaws. These transport not only people, but also little surpluses of vegetables or chickens, so that rural people can capitalize on minor surpluses at the local small town, in a way not previously known. Indeed, one could almost say that the greatest transformation of rural India ever is now occurring, because of rural transport, and occurring fastest where the greatest number of villages are on all-weather roads.

One last point needs to be made about this region: in all three census periods shown in the Tables it has had the highest population growth rate. It has also had (together with the west) the highest per capita income growth. In the short term at least this confounds a Malthusian view of India. Whether it is storing up trouble for the future if the economy slows is another question.

## Gujarat and Maharashtra

The second High Speed region includes most of Gujarat and Western Maharashtra. They are among the most urbanized areas of India, with more than 30 per cent of the population living in urban areas. Industry in Gujarat has centred around the cotton textile mills of Ahmedabad, but the state has also been the site of considerable inward investment in chemicals. Maharashtra is now India's pre-eminent industrial state, having surpassed West Bengal. Virtually all the activity is centred around a cluster of cities in the Bombay region – by which is meant a territory which includes Pune

(Poona) high on the Western Ghats, Thane – Bombay's new satellite city – Ulhasnagar, and of course Bombay itself. Though Bombay's fortunes were based on its magnificent harbour and a history of overseas trade, and also on cotton textiles, the industrial diversification since independence has been remarkable. It is now a centre for the petrochemical industry – being the land centre for India's off-shore oil industry. It has major motor vehicle plants, and investments in heavy and light electricals. To complete the high-tech galaxy, the region also boasts one of India's nuclear generating plants.

The rural areas of these states have also progressed well, but not to the same extent as Punjab and Haryana. In Gujarat cotton and wheat yields and production have grown. But agriculture is still considerably dependent on coarse grains like bajra (a millet), and still has inadequate irrigation development. To some extent this inadequacy may be rectified by the fact that Gujarat is one of the three states that will benefit from the new Narmada dams and irrigation schemes – though these are on such a massive scale that development will take many decades to come to complete fruition, and might not occur at all because of the environmentalist backlash. In June 1990 the World Bank took the unusual step of inviting outside consultants to review again the environmental costs of the schemes.

Rural Maharashtra includes the wet coastal rice area, and the drier Deccan high plateaux. The latter in particular have areas of poverty associated with the drier and more erratic climate and the attendant more precarious agriculture. Given that Maharashtra is the third most populous state of the Union, it is not surprising that it figures fairly highly in the list of states with high totals of poverty: it is in fact fifth in that list, despite being part of 'High Speed India'.

## The south

This is a disparate region too: it includes not only the sleepiest of the giant Metropoli of India – Madras – but also the extremely dynamic Bangalore, capital of Karnataka, and a centre of some of India's high-tech industry, including aeronautics and some of its finest electrical and chemical industries. It also includes an arc of towns stretching south-west from Madras towards Coimbatore, which have had a long history in textile production. Of the High Speed regions, this one has the lowest per capita income growth rate

(1.3 per cent 1961–81 compared with 1.5 per cent for the other two), but perhaps this is not too significant a gap. Part of this may be explained by the nature of industry – far more of it is in domestic and handcraft production than in the other High Speed regions, and commensurately less of it in the large-scale 'modern' sector. The value-added per industrial worker is significantly less than in the other two High Speed regions.

It is also true that the basis of agriculture in this area for many is rice, and rice has done less well in the Green Revolution than wheat. Having said that, this region has been notably more successful in adopting new rice varieties than Low Speed India, in the east. Part of the reason for this is that the pattern of urbanization has been more widespread, with more smaller towns. This then links, as in the north-west, to a better pattern of infrastructure, and the possibility of a better distribution of inputs and marketing of surplus. In both rural electri-fication and road connectivity, this part of India is very much in company with the High Speed zones, not the slow east (see Table 2.3).

## LOW SPEED INDIA

Low Speed India includes Uttar Pradesh (it ought not to include parts of the west of the state, but for statistical purposes they have been included here), Bihar, West Bengal, all of which can be thought of as parts of the humid and densely settled Ganges plains, and Orissa and Madhya Pradesh – the northern and eastern parts of the Deccan, not so densely settled. In many ways this area is the victim of its own historical success, in that the plains areas are the densest settled areas of India, having been able through simple agriculture from time immemorial to sustain a large peasantry. Land holdings per farmer in many areas are much less than half that of Punjab, so it would appear prima facie that there is less chance to accumulate capital. But there are also many compounding social factors: the history of land tenancy from colonial times till Independence in 1947 led to stronger exploitative landlordism than elsewhere in South Asia, and to a greater number of landless share-croppers and landless labourers. The worst excesses have been eliminated since Independence, strengthening the middle peasantry, but the problems for the landless are still acute. This is also the region with the highest number of scheduled castes, that is to say the groups who were once known as untouchables, and ritually excluded from many areas of life.

25

This social structure has undoubtedly inhibited capital accumulation and investment over a long time period. It has also done so in recent decades. The net effect is that, although this region has the lowest population growth rate of any of the regions, it also has seen no economic growth per capita at all in the last thirty years. It still has the lowest level of urbanization, and the lowest level of rural infrastructure, so that there is neither the opportunity nor the market incentive for rural areas to commercialize in the same way as in other parts of India.

It has also suffered from political change. Calcutta was capital for much of the British period, but then that status was transferred to Delhi in 1911. It used to be the metropolitan heart for the whole of Bengal, processing jute from the eastern half. But in 1947 the eastern half became East Pakistan, and in the turmoil the city's population was expanded by an influx of several million refugees (the same also happened to Delhi), and two years later a trade-war developed, which effectively closed the border and cut the city from the hinterland. Although the jute industry has survived, it has not expanded, as the product has become increasingly less significant in international trade.

Some of this tale of stress and distress seems inappropriate for the most mineral-rich area of India. In the north-east Deccan area of Chotanagpur – roughly corresponding with the Deccan parts of Bihar state, and in the adjacent parts of West Bengal, lie the overwhelming bulk of India's mineral resources – coal, iron ore, bauxite, copper ores and even uranium. Much of this is in areas which were not central parts of older Indian civilizations, and were not given much attention by the British colonial state either. Only coal, necessary for the railways, was developed much before Independence, along with the first indigenous steel plant at Jamshedpur, dating from the First World War. (The city is named after Jamshedi Tata, founder of the Tata group, still one of India's dominant industrial combines.)

In the forty or so years since Independence, India has planned a massive industrial expansion to catch up on these years of neglect. Much of the heavy industrial investment in iron and steel, in coal, and in heavy engineering has been sponsored directly by the central government in this area, and a significant number of new industrial cities have been founded. The question is, why has this not had a greater impact on the region in general, than it appears to have done?

The answers are not easy to find: to some the process has been one of internal colonialism, with the development of an enclave economy. Many of these new towns are inhabited by a significant number of immigrants from non-local areas. The finance comes from exterior regions, and the produce is shipped out of the region too. Even local locational advantage has been negated by pricing schemes which have charged flat-rate prices for output throughout India. In addition, despite the market power that these new cities could represent for the rural hinterlands, the latter have had little investment in the roads and road transport to enable them to benefit, and since the rural regions are poor, there has in any case been little capital available to make the investments to supply the towns. Even perversely, much of the little local capital that is available has gone into urban services, such as small hotels and bars, rather than into productive assets in the rural areas. None of this is to say that the local income multipliers will not one day prove dynamic and valuable to this lagging region: but it is to say that so far the take-off has not happened.

## 'SPECIAL' NORTH-EAST

The last region of India has historically rarely been subjected to central powers. The British incorporated much of the Brahmaputra Valley and the hills around, into India in the latter part of the nineteenth century, and it has been in a sense 'bequeathed' to the modern Republic. Local cultural traditions are usually not Hindu, and may of the populace have a basically tribal organization. Many have been the subject of Christian missionary activity, and have an education and cultural orientation different from the people of the Ganges heartland. The area does have some resources – a small oil field in Assam, and tea plantations in the hills, needed for India's thirsty tea-drinkers if not for the world's. Settlement density has been low, so that here in the Vale of Assam there has been land to spare – an unheard-of commodity in so many other parts. This has lead to a continuous stream, if not wave, of immigration, much of it illegal (from Bangladesh), and consequent clashes between local people and the newcomers. In Assam data are less readily available than for other areas of India, since so often census collection has been suspended because of local political trouble. Armed insurgency has been a fact of life for several decades in Nagaland and Mizoram.

The central government has responded by awarding statehood to units which are much smaller than elsewhere in India, and by per capita levels of state expenditure which have been well above average. Mizoram and Manipur have the highest jump in urbanization between 1961 and 1981 of any areas of India, an inevitable response to insurgent warfare and high levels of direct expenditure and indirect expenditure by the military.

## ENVIRONMENTAL ISSUES

Modern rhetoric about the environment often portrays India as in the grip of a fundamental crisis. Trees are supposedly being stripped from every last vestige of forest, the mountains being eroded, the plains lost to salinization, the cities suffocating in their own gaseous effluent. The great difficulty is to separate out inevitable and natural processes from degradation unnecessarily accelerated by humans, and to understand the true extent of any problem which might be acutely manifest in some selected region. Above all it is important to remember that this is naturally a very dynamic landscape in many areas. Except in the hard rock areas of the Deccan, it is a landscape still being built by ongoing processes. The northern mountains are still undergoing upthrust – at the rate of about 6 cm a year (Tapponnier et al. 1986: 116; Ives and Messerli 1989: 98) – so in very crude terms if they are eroding at 6 cm a year – which is a massive amount – they would be in some sort of 'equilibrium'. The plains beneath them are after all made of recent fluvial and other sedimentary deposits, and the delta in Bengal is the most active on earth. It deposits silt 2,000 nautical miles out into the bay of Bengal (King 1983: 119). Large rivers have captured others in the Punjab in the last three centuries. The Kosi descending from the Himalayas of Nepal into Bihar has built a fan across which it has moved 80 miles west in the last 200 years. These great shifts affect the local hydrology. Fatehpur Sikri, the fabled ghost capital of the Moghuls built south of Agra, had to be abandoned when the local water supplies suddenly dried up. The consequences of India's dynamic environment are an indelible part of India's history.

### Land degradation

Land degradation invariably means losses to productive farmland, although it can and should also mean losses to forest soils such

28

that only degenerate forest can be re-established. Losses to productive farmland can come about through removal – i.e. erosion or excavation – and changes in chemical composition, brought about by overuse and impoverishment, salinization or by chemical adulteration.

Erosion is undoubtedly severe in some areas, and particularly so in some hill areas. The problem in understanding this is that the natural rates of erosion are also very high, particularly in mountain regions such as the Himalayas. Detailed studies suggest that terraced mountain land is in fact better able to retain water, and gives rise to less soil removal, than equivalent forest areas. When terraces do collapse, they do so in a spectacular way, domino-like in the sequence of collapse down a hill-side, and give rise to fears that massive erosion is taking place. But they are usually reinstated, and the original scar is less visible after a few years. In so far as some soil is lost, these slips are in a sense a minor compensation for erosion that has not taken place because of the terracing. What is certain is that unterraced cleared forest does cause a great increase in runoff and erosion, and is extremely damaging.

Erosion in gullies and ravines reaches spectacular extent in the soft alluvium of the Chambal basin, in the area straddling the borders of south-western Uttar Pradesh, north-western Madhya Pradesh and eastern Rajasthan (Centre for Science and Environment 1985). This area has quite obviously at various times in the Pleistocene suffered erosion. The question now is whether it is suffering accelerated erosion – and undoubtedly there are areas where this is probably the case. Figures are quoted for villages deserted in this district and that – yet India's history is one of shifting settlement as local conditions change. But it is not necessarily the case that all instances are the result of human interference with the local ecology. This area is adjacent to the Aravallis of Rajasthan, around which have accumulated at some time in the past sand dunes which mark an earlier more extensive margin to a larger Thar desert. These dunes have been fossilized and vegetated. Some now have their vegetation degraded by grazing and fuel collection, and have become mobile again in some parts. But it is also possible that an increasing incidence of drought which might be associated with a 'normal' pattern of climatic change is beginning to make the area sensitive again to erosion. Disputes about causes should not however mask the fact that in some places remedial action can be worthwhile and beneficial.

Direct excavation of soil for brick-pits is common in extensive areas at the margins of any medium or large city, and even in rural areas too. These pits are usually shallow, dug by hand to feed small kilns. They are therefore areally more extensive than the kinds of commercial pits seen in the English clay lands. In effect urbanization is vastly more damaging and extensive than the 'built-up' area suggests, because of the 'built-down' result of the production of these building materials. The reason for this extensive development of a small-scale industry is to be found in the relatively higher costs of transport within the economy. Neither the lorries nor the roads make long-distance conveyance of high-bulk low-value commodities economically competitive while licences for pits can be obtained fairly easily locally. In the river lowlands and plains bricks are also needed because of the lack of stone, for hardcore for roads and concrete. Bricks made by hand and then pulverized by hand are the only available material, meaning that even in rural areas as roads extend, so do the brick-pits.

Soil can also be damaged by chemical changes. In the Ganges plains of Haryana and Uttar Pradesh extensive irrigation systems have contributed to a steadily rising water table – in places coming nearer enough to the surface to allow salinization of the soil to occur – the land becomes 'usar' land. Satellite imagery is beginning to reveal what is obvious at ground level, but more difficult to quantify – that very large areas are now badly affected. The cause in a sense goes long back – since canals started in the 1850s. But in recent years the density of the canals and the months of the year when they are used have both been extended. Soil can be badly affected by irrigation schemes where surveys at the planning stage did not discover inappropriate local conditions or unsuspected difficulties. The world's longest canal built for the world's largest canal scheme, now known as the Indira Gandhi Canal, under construction in Rajasthan since the 1960s and still a long way from being finished, has begun in places to irrigate land where there is a hard-pan previously unsuspected between 5 and 20 metres below the surface.The resultant waterlogging can be imagined.

Lastly, chemical fertilizers are increasingly used to stimulate crop production. The use of fertilizers in all states has increased significantly, and is one of the reasons why India has been able to keep food output in step with population. But usage is normally only of nitrogen, phosphorus and potassium. The effect of these in long-term use on tropical and sub-tropical soils is not really well understood. One suspected affect is that other trace-elements – micro-nutrients – become removed, and yields and crop quality become badly affected. The commonest complaint is about the removal of zinc. Restitution of such trace elements is a complicated and delicate business, and the country has too few soil scientists and too few farmers with this kind of knowledge to be able to handle the problem properly. Thus the primary intervention which appears beneficial may well lead to longer-term side-effects which are much harder to control.

## Forests

The shrinking of India's forests has become not only a national (Khoshoo 1986) but also an international *cause célèbre*. There are some areas of denudation which are almost welcomed by environmentalist groups because of the powerful message which they deliver – around the famous Doon valley near Dehra Dun in the foothills of Uttar Pradesh for example. What no one knows is whether these highly visible areas of clearance near roads and centres of population are repeated elsewhere. Some authorities believe that in fact there is little evidence that there is extensive deforestation. Ironically, it seems that the best way to preserve the forests is to avoid building roads to rural communities, and to avoid bringing electricity. The latter can power saw mills, and the former provide the export routes for timber.

As well as from cutting for commercial timber, the forests are supposed to suffer from the extraction of fuelwood and the grazing of cattle. Firewood still provides at least half of all India's domestic cooking fuel – including for major urban areas. Per therm of generated heat it is more expensive than bottle-gas, yet it is the fuel of the poor, who cannot afford stoves and the deposits on gas-bottles, and are unable to relate to the gas distribution systems. Other fuels in urban areas include dung, bought from urban dairies which keep stall-fed cattle in town, coal dust, coal, rubbish and kerosene. None of these has proved as attractive and easy to handle as wood for the masses of poor. It is not certain, however, where all the wood and charcoal comes from. It is assumed that much of it is long-distance and 'leaks' out of state forests. Indeed in Delhi 20 per cent of the railway wagons arriving at Tughlukabad station are for wood fuel, mostly coming from Madhya Pradesh. But a study of Hyderabad (Alam et al. 1985) has shown that at least half of the wood has come from private land near the city, not the state forests at greater distances. Partly this is the result of harvesting dead trees that generally occur around villages and at field boundaries: but increasingly this is the result of the planting and harvesting of eucalyptus in private plantations. The acreage under this tree has grown significantly. It is fast-growing, drought tolerant, and tolerant of divergent soil conditions. It has however a bad ecological reputation – being blamed for declining water tables, a leaf litter that harms the soil (as rhododendron leaf litter poisons the soil for many organisms), and a leaf which is unacceptable as an animal fodder (another popular new tree leucaena does have considerable although problematical value as fodder.) The extent of these criticisms is not the subject of scientific certainty. What is clear is that it can make money, as fast if not faster than other crops, and that while it does so, it is relieving pressure on alternative forest sources. It may not benefit the poor and landless, but its very real success in the countryside is testimony of something.

Official statistics on forest cover are confusing. Those collected at district level often refer to land classified as forest but no more wooded than is Dartmoor Forest in Britain. Again satellite imagery is beginning to reveal a slightly clearer picture. It is clear that good quality forest has receded rapidly in the last thirty years, and that there is really little now left for a country once know for its jungles and jungle animals. Some of the remnants are now increasingly well guarded, and many have been designated sanctuaries under such schemes as 'Operation Tiger' which seems successfully to have protected the remnant tiger populations.


## Water

Water is used more for agriculture than for urban and industrial use – not surprisingly in a country which is so dependent on irrigated agriculture. Only 8 per cent of total water use is currently for urban and industrial purposes. But the latter demand is increasing annually in amount, and since much of it is returned to water courses in polluted state, the impact of this misuse is also mounting. The two effects combine, as in the Ganges, where there is increasing demand for water for irrigation, reducing in particular the dry season flow of the river, and yet an increasing use of the river by urban areas for both intake and discharge. At Kanpur and at Varanasi (Khoshoo 1986; Singh et al. 1988) the river has achieved extremely high pollution levels. This has led to a major campaign to clean up India's holiest river, involving amongst others Thames Water as consultants, in the belief that they have proved successful in cleaning up the Thames. The investment required in effluent treatment works at industrial sites is enormous – as also the need to invest in proper sewage treatment works. The demands on the river are already such that India, which diverts water down the Hooghly past Calcutta from the Farakka barrage during the dry season, in consequence no longer lets enough water downstream for Bangladesh's demands, and the issue has escalated to an international dispute. Yet what we are discussing is the area with the greatest number of India's poor, and the least urbanization. As, hopefully, development proceeds to allow the citizenry a better standard of living, and as this implies increasingly large cities, the future for water management in the whole basin looks critical.

The same problems are replicated over the whole of the Deccan, in river basins which are even more seasonal than the Ganges. The water supply for many urban areas is already precarious: Madras has had to have additional water freighted in on tanker trains in several recent years, for distribution via water carts. It is, incidentally, salutary to realize that these water shortages mean that in many urban areas the idea of flush sanitation is not just economically impracticable but also technically infeasible (as many as half of the population excrete in open waste land, lacking proper sanitation). In no area is the record on pollution control good, nor the imposition of legal standards satisfactory. The stories of rural communities finding their rivers poisoned are too common to be smoke without fire.

## Air

India's airspeed for much of the year is very low, meaning that what goes up in one place, tends to come down again in the same place. The emissions in urban areas are bad, and getting worse. Not only has vehicular traffic increased in all urban areas dramatically over the last thirty years, but also the standards of exhaust control are extremely low. Industry is a major polluter too, with few smoke-stacks having proper scrubbing equipment. But worst of all is the domestic contribution – not from fires for heating as used to be the case in London, but from fires for cooking in the morning and evening. The inferior coal used and the wood, the oily rags, leaves, cardboard boxes and dung cakes, put out a heavy pall of smoke. The particulate matter air pollution in the big cities is now some of the worst on earth, and contributes to high incidences of pulmonary disorders. Power-station output is also a significant source of sulphur oxides, contributing to acid rain in the wet seasons – though the extent of this as a problem is not really known, since soil chemistry in India is not the same as for example on the hard-rock shield of Sweden. A recent survey of the inhabitants of Varanasi (Singh et al. 1988) has shown that they know that it is a problem, that it can be harmful, but it also shows that in all groups except the highest income groups, it is far less of a problem than getting a job, buying food or getting accommodation. The message seems to be that the public accepts that cleaning up the environment is something that comes after other necessities have been catered for – a pattern which the West has also followed historically, though not at a time when a modern chemical and plastics industry had been developed. It is also an attitude from which big business seems happy to profit. Whether the scale of damage will ultimately have a negative impact on economic growth itself is questionable. Whether the pressures will ever succeed in bringing polluters to book is also questionable. Five years on, the victims of Bhopal, where a gas leak from a pesticide factory killed thousands, are still awaiting compensation, and the legal administration of the case has become a political toy at the highest national levels.

## Irrigation

Irrigation has been of growing importance in India since Independence in 1947. Irrigation is seen as a panacea – insulating agriculture and food supply from the vagaries of the weather. It thereby enhances the attractiveness of new technologies, new crop varieties and fertilizer. Indeed the percentage of cropped land under irrigation has risen between 1960 and 1980 from somewhere around 18 per cent to somewhere around 32 per cent, while the gross cultivated acreage has grown from 132 million hectares to 174 million hectares (Muthiah 1987). (This according to Farmer 1974 must be somewhere near the limit of agricultural colonization.) The full irrigated potential for India is quoted at about 110 million hectares – though it is hard to reconcile this very high implied overall percentage figure for irrigation with the large tracts that are likely to remain rain-fed only.

Irrigation comes in many forms: large-scale dams with canals – officially large-scale means with a command area of more than 10,000 hectares; medium-scale dams, local traditional earthen dams known as tanks, and tube-wells, drilled by modern rigs, and open wells constructed in a traditional style. The wells may be energized by human or other animal power, or by diesel or electric pumps.

There have been 1,040 large or medium-scale schemes since 1947. The full potential for dams and canals of much of the Deccan has been completed. The remaining large-scale dam sites are mostly within the Ganges basin, whose largest tributaries descend from the Himalayas. Whether the grandiose schemes for these rivers will ever be completed depends not only on international co-operation, particularly with Nepal, but also on the strength of the environmental lobby that knows that such dams are potentially dangerous in a seismicly active area, that they will silt up quickly, and that canals contribute to waterlogging and rising water tables. But the dam-building contractors' lobby is wealthy and powerful. It seems likely that much of the great scheme for the Narmada valley, the last great untamed Deccan river, will go ahead.

Though these schemes grab the headlines, it has in fact been well irrigation that has exploded everywhere. Large and medium canal schemes account for 39 per cent of net irrigated area, wells and tube-wells for 46 per cent. (The remainder is mostly local tanks.) Even in Punjab, mentally identified with canal schemes, wells now account for 61 per cent of the irrigated acreage. The number of installed tube-wells has risen from 0.2 million in 1960 to 4 million now – and the number of diesel sets has shown an almost exactly similar change. The number of electric pumps has shown a similar rise – which might sound like duplication. It is indeed. Power supplies are so uncertain that although electricity is cheaper than diesel, few farmers can afford to rely on it alone. So diesel, expensive and polluting, is heavily used as a stand-by.

The wells in most areas are seen as environmentally sound. They promote downward movement of water in the soil, thereby avoiding salinization. They use local water, and do not import supplies from other regions, again avoiding contributing to waterlogging – though in some areas of the Deccan powered wells are clearly mining water at a faster rate than replenishment, and water tables are sinking, to the detriment mostly of the poorer and smaller farmers. In many areas the catchword now is 'conjunctive' use – meaning the use of tube-wells and canals and other sources together. Canals contribute to farms within the command area, and to ground water. The wells guarantee local topping-up of supplies, and reduce waterlogging. The 'theory' is in fact more like an a posteriori explanation for something the farmers have sorted out for themselves.

## RELIGIOUS AND REGIONAL TENSION

India is overwhelmingly Hindu, although that term embraces a multitude of castes and beliefs, and is more a cultural than a religious statement. But whether or not one describes Hinduism as a religion in the Western sense, it is certain that at least 80 per cent of the population does not profess a faith which is more likely to be covered by the term religion, such as Islam, or Sikhism. The status of being Hindu is conferred by birth – one cannot (despite the stream of Western 'hippies' who might believe otherwise) be converted to Hinduism, since caste membership at birth, part of the cycle of reincarnation, is an essential concomitant. But other religions in India can be evangelical – be it Christianity, Sikhism, Islam or even versions of Buddhism. All of these religious groups have shown faster rates of growth than the mainstream Hindus. Partly this can be explained by demographic factors rather than confessional ones – it seems that birth rates in the Muslim community are higher than those in the Hindu. Though there is a greater objection to birth control by Muslims, the fact that proportionately more Muslims are amongst the poorest sections of society may mean that it is poverty not religion that engenders the higher growth rate. Conversion is also of some importance, particularly in the south, where the evangelical crusade can provoke a backlash from the Hindu majority. With Buddhism, clearly the massive growth from a small base cannot be explained by demographic factors. Conversion has been a very strong factor in recent years, particularly in Maharashtra, where the 'untouchable' castes have been converted, often en masse – just as lower castes were attracted to Islam en masse in Bengal many centuries ago.

Hindu domination of India is not threatened: but it is seen to be threatened. India, a secular state which has striven to make itself a home for all creeds, distinctive in this sense from Pakistan and Bangladesh which have religiously inspired constitutional status, has yet found that religion has been a growing factor in politics. Communal riots have always been a feature of the political landscape of some states, but of recent years they have almost reached endemic proportions in Gujarat, and in the last year have developed into alarming features of political activity in the Gangetic heartland of India, where struggles over a Muslim mosque built centuries ago at Ayodhya on the putative birth-place of Lord Rama, one of the Hindu incarnations of God, have reached national symbolic status. This struggle and the new Hindu fundamentalism led to the collapse of the government in 1990.

At the same time, struggles against Delhi's domination have broken out in Kashmir, the only state in India with a Muslim majority. The exact political status of the state has been in dispute with Pakistan since 1947, although since 1949 the larger and more populous part of the former princely state has been incorporated within India, and for some of that time has seemed to be able to operate democratically within that framework. India blames Pakistan for the current trouble, although there is no doubt that the insensitive security crackdown which started in 1989 has alienated the majority of the population.

Taken together, these two developments look ominous for the future unity of India. They seem even more ominous when one remembers that in Assam and other states of the north-east, again not part of mainstream Hindu India, there is again open rebellion and examples of guerrilla campaigns, and that the Tamil problem of Sri Lanka has dragged India into involvement in another destabilizing communal conflict.

The extent to which these centrifugal forces are pulling India apart is the subject of endless speculation. There is no doubt that the central government is now weaker in relation to the states than it was in 1947, but that is partly because as the local economies have grown and become more complex, and as local possibilities, particularly in High Speed India, for revenue raising have improved, then it is an inevitable and proper development for local power to increase; it does not necessarily mean secession. But politics at the central level has increasingly begun to hang on coalitions of parties which lack the kind of pan-national appeal that Congress once had – and this is a development that worries some observers. The counter-argument is that there is a large and influential business and academic community, which shares a common language – English – and which benefits from the large common market that India represents. This group has no wish for disintegration: but politically it has lost power as Congress has fractured and the new popularist politics of community have taken off. It is a worrying development.

While increasingly it is communal issues that dominates politics, the kinds of class alignments and associated ideologies with which the West is familiar seem less and less relevant to political manifestos in India. The rhetoric and to some extent the reality of Congress policy used to be clear: socialist control of the 'commanding heights' within a mixed economy, overseas non-alignment, but yet with a treaty of friendship with the USSR and extensive East bloc trade. Now all this has been melting in favour of a more open economy, partly at the behest of the IMF, in order to stimulate competition, and to catch-up rather than fall behind in the international technological race. It fits with the world fashion of the 1990s. Yet in Bengal and in Kerala communist governments have or are trying to assist some of the weaker sections of society but within the confines of the electoral system. A creeping political pluralism, if not anarchy, seems to be accompanying the drive for a more open and competitive economic system – and given India's historical propensity for particularly social groups to monopolize particular economic functions, one should not necessarily expect that the new openness will lead to a more competitive and economically rational system.

The parallels with Europe are often very instructive. At birth in 1947 the modern republic of India had a unified currency, a strong central government, a unitary defence policy, a single external customs barrier, and constituent states whose power derived from devolution rather than from gradual surrender of the original sovereign rights of acceding independent states. In theory this is all still the case. But now that the states have been realigned with linguistics nationalities (since 1956), there is little chance of the centre denying any of them their basic statehood. Increasingly financial power is devolved. Some even have nascent independent foreign policies, in the sense that they will negotiate with major sources of external credit such as the World Bank for their own projects. Internal barriers to the free movements of goods exist, sometimes at central government behest as in the case of food, sometimes at the state governments' behest because of differences in taxation policy. The major arena in which this devolution is almost completely absent is in the armed forces.

In the case of Europe, the move towards federation, a common currency, a common defence force, is very slow and faltering, slowly it is building up from a base, from nations which in many ways are economically more equal, culturally and linguistically closer, than the states of India. There should be little surprise that in India's case there is a reverse process, trying still to accommodate the end of Empire.

## EXTERNAL RELATIONS

Trade within the whole of South Asia was open until 1947, and with the rest of the world, particularly the Commonwealth, was also fairly open. But the barriers against both were erected rapidly after 1947. In the case of internal South Asian trade, the political (and military) war with Pakistan erupted into a complete trade embargo from 1949. Indian coal was no longer exported to Pakistan, (East) Pakistani jute and (West) Pakistani cotton no longer exported to India. These deficiencies were fairly swiftly made good by domestic substitution. Tariff and quota barriers were also erected against external industrial goods, as India sought to expand its own domestic industry. India became the world's tenth ranked industrial power, but much of it was at high costs and in world terms inefficient. The policy probably reached its zenith in the 1960s, and by the 1970s questions over its value were already being raised. After the oil shock of 1973, India was forced to take stock. The appraisal showed the extent to which, during the process of concentrating on substituting for imports, traditional exports had declined, and though there were some new manufactured exports these had not made good the deficiency. Between 1960 and 1985 India's share of world exports declined from 1.0 per cent (already a low figure) to 0.4 per cent (Muthiah 1987: 254). This relative fraction is put into sharp focus by comparing the absolute value of India's exports in 1985, $7.9 billion, with those of Singapore, $22.8 billion, and of Hong Kong, $30.2 billion. Where there were export booms, in hand-loom cottons etc., many were serendipitous. The policy of state control over trade had also produced some new barter deals with states such as the Soviet Union, but which signally failed (except in the case of military arms) to provide India with any hope of technological advance. With a faltering balance of payments, India's advances to the IMF secured loans on the same kinds of restructuring conditions which have applied to many other Third World states. The consequence has been since the mid-1980s a slowly growing openness, a slow dismantling of barriers, more international joint ventures, and the exposure of many sectors of industry to some chill winds. Pressure groups in India have tried to alleviate the worst: there is an official category of Sick Industry, which can benefit from specific kinds of government help.

The new trade is in the main with the West (including a very major Japanese invasion). It has had very little to do with the rest of South Asia – which from India's point of view is in many instances more backward.

The external political relations of India have reflected a considerable consistency since Independence. Nehru vehemently wanted to avoid a neo-colonialist future – and that was what lay behind the drive for autarkic industrialization as much as anything else. The United States could have been its ideological ally – uniting the world's largest democracy with the most powerful. But the USA has seen its external allies in terms of their usefulness in the containment of communism, and has only secondarily been interested in their internal ideology, which in India's case was anyway suspect, because of Nehru's openly socialist rhetoric. India did not join CENTO (Central Treaty Organization: originally the Baghdad Pact), as a bulwark against communism. Pakistan on the other hand did, and since Pakistan's relationship with India has been wholly antipathetic (mostly because of the dispute over Kashmir), from Delhi the emergence of a Washington–Karachi(Islamabad) axis was viewed with alarm. When China and the USSR also split, and China and Pakistan formed close alliances, India naturally turned to the balancing power of the Soviet Union, leading to the trade ties we have mentioned, military assistance and a treaty of friendship signed in 1971. Despite being humbled by the Chinese in a border war in 1962, none of India's neighbours now would doubt her regional dominance. In 1970–1 the East Pakistan crisis spilled over into India as refugees fled from East to West Bengal. When war broke out, India took a prominent role in the military defeat of Pakistan in the east, and the establishment of Bangladesh. She has also flexed her muscles, to indeterminate effect in a military action in Sri Lanka and to certain military effect in the Maldives, and has acted with considerable resolve in disputes with Nepal. Sikkim has been completely absorbed in India, and Bhutan retained within a protective framework. She has also supported the communist government of Afghanistan since the Russian invasion, another point of tension with Pakistan. The curiosity is that this undoubted domination has not lead to any real opening of other ties.

In 1979 the South Asian Association for Regional Co-operation was formed, at the instigation of President Zia of Bangladesh. This is a nascent movement for closer regional integration, but it has very insubstantial terms of reference. Only multilateral issues can be discussed. No bilateral issues can be brought to the meetings: so issues involving India with one of her neighbours alone are not part of the agenda. This is to stop the smaller countries ganging up on India, but it effectively emasculates much real progress. The pattern has been repeated in specific cases. Even relations up and down the Ganges Valley have been treated bilaterally. India will negotiate downstream with Bangladesh and upstream with Nepal: it will not agree to trilateral talks over this major international resource.

## THE GEOGRAPHY OF IGNORANCE

Any government official will agree that there is an overwhelming ignorance (though the magnitude of the unknown can be fairly well estimated) about the 'black' economy – that is the economy that operates on undeclared financial transactions. It is not simply a matter of concern over lost taxes. The size of the black economy is such that many policy initiatives of the government can be frustrated by counter-movements in the counter-economy. It has its physical manifestation too – both in the massive and unregulated squatter settlements, and in the splendid residential developments of the rich, rich beyond known means. Indeed many would argue that speculation in urban land is a major source of this wealth, and that it can subvert any urban planning policies – since lowly paid officials are always susceptible.

It is also often the case in the country that the ostensible power structure is not the real one. As an example, the case of irrigation has been documented. The scale of corruption over water alloca-tions is attested by the fact that an official may pay ten times his annual salary to occupy a job for just two years – and he can come away from that posting with a handsome profit. In a sense this is not complete ignorance – the problem is recognized. But no one knows whether to rectify the situation by openly accepting these market forces and working with them, or whether to attempt to enforce the behavioural norms which have been bequeathed to the country from colonial times. (The British officials were far less corruptible: but they paid themselves salaries which in real terms were orders of magnitude higher than the inflation-eroded 'socialist-egalitarian' salaries of modern official India.) There are of course powerful vested interests that profit most from the present duplicity and which will seek to perpetuate it.

A similar but different duplicity affects many official statistics. Officials asked to report statistics to their superiors commonly take the most optimistic line. The number of schools may be cited – but these are often incomplete building shells (the contractors have made off with the balance of the money), whose staff are more absent than present, and many of whose 'pupils' never attend. The number of new drinking water pumps installed may be known, but not the number which are actually working. The number of villages connected to the electricity supply may be known, but not the actual number of hours of uninterrupted supply. Even the number of consumers is not known: the theft of electricity by illegal step-down transformers hooked on to the grid means that in Calcutta a quarter of the current is unaccounted for.

In the countryside the real extent of soil erosion and deforestation is not known. Local statistics can be quoted, but to extrapolate to the all-Union level would be very dangerous. The population numbers are known fairly well, but little is known about the dynamics of migration and the flow of remittances. Something is known from sample surveys about fertility and about death rates, but there is no Registrar of Births, Deaths and Marriages. In many states only land transfers are recorded, and then only some of the transactions – and there is no proper land registry. So enforcement of land ceilings, which can in any case be circumvented in many ways, is almost impossible. In other words, the state's recording and understanding of the actual on-the-ground local demography and economy, as opposed to the aggregate, is weak.

Above all the greatest ignorance is one of knowing what are the truly valid policy options. India is still overwhelmingly rural, its rural population often abjectly poor. Its cities are growing fast, fuelled by a newly dynamic industrialization, but which yet employs few people directly. The cost of this urbanization is very high, it monopolizes much of the capital – and particularly 'black' capital. Is there a future in which rising living standards, increased employment opportunity and job security, can be provided either in villages or in new small towns integrated with their local rural communities? Is there a Gandhian future possible, without the excesses of pollution and squalor commonly seen in the great cities of the present? There is a stream in Indian political thought which proclaims such a future, and the virtues of appropriate technology, and of ecologically sensitive sustainable development. But the ignorance is precisely that no one knows whether or not it is a dream, and if not what are the real means, as opposed to mere exhortation, which could achieve it. So far the cynics and free marketeers are in the vanguard: and they have before them the alluring model of the wealth that urban-industrial technology has given the West.

## REFERENCES

Alam, M., Dunkerley, J., Gopi, K.N., Ramsay, W. and Davis, E. (1985) *Fuelwood in Urban Markets: A Case Study of Hyderabad*, New Delhi: Concept Publishing.

Bradnock, R.W. (1990) *India's Foreign Policy since 1971*, London: Royal Institute of International Affairs/Pinter.

Centre for Science and Environment (1985) *The State of India's Environment 1984–85: The Second Citizens Report*, New Delhi: Ravi Chopra Ambassador Press.

Farmer, B.H. (1974) *Agricultural Colonization in India since Independence*, London: Oxford University Press, for the Royal Institute of Strategic Affairs.

Ives, J.D. (1989) 'Deforestation in the Himalayas: the cause of increased flooding in Bangladesh and northern India?', *Land Use Policy*, 6(3): 187–93).

Ives, J.D. and Messerli, B. (1989) *The Himalayan Dilemma: Reconciling Development and Conservation*, London and New York: The United Nations University and Routledge.

Johnson, B.L.C. (1979) *India: Resources and Development*, London: Heinemann.

Khoshoo, T.N. (1986) *Environmental Priorities in India and Sustainable Development*, New Delhi: Presidential Address, India Science Congress Association.

King, L.C. (1983) *Wandering Continents and Spreading Sea Floors on an Expanding Earth*, Chichester: John Wiley.

Muthiah, S. (ed.) (1987) *A Social and Economic Atlas of India*, New Delhi: Oxford.

National Institute of Urban Affairs (1988) *The State of India's Urbanization*, New Delhi: NIUA.

Singh, O., Kumra, V.K. and Singh, J. (1988) *India's Urban Environment*, Varanasi: Tara Book Agency.

Tapponnier, P., Peltzer, G. and Armijo, R. (1986) 'On the mechanics of the collision between India and Asia', in Coward, M.P. and Ries, A.C. (eds) *Collision Tectonics*, Geological Society Special Publication no. 19, Oxford: Blackwell Scientific.

Tata Services Ltd (1989) *Statistical Outline of India 1989–1990*, Bombay: D.R. Pendse for Tata Services Ltd.

# 3

# THE CHANGING GEOGRAPHY OF THE STATES OF THE SOUTH ASIAN PERIPHERY

*Robert W. Bradnock*

## INTRODUCTION

With India at its core, South Asia could be perceived as a classic example of a centre-periphery region. Marginalized geographically and politically from the dominant regional heartland, throughout the last twenty-five years the countries of the periphery have struggled to develop a base from which to exercise effective sovereignty over their national territories. From Pakistan in the west to Bangladesh in the east, from Nepal and Bhutan in the north to Sri Lanka and the Maldives in the south, India's neighbours show a wide range of strongly individualistic characteristics (see Figure 3.1). Yet all share the overwhelming importance of their proximity to and relationship with India. All have far stronger historic and cultural links with India than with their other neighbours, tying them into a South Asian cultural region, however much they might wish to identify themselves with an alternative. This importance has persisted throughout a quarter century of dramatic changes in the region, a period during which none of the countries, except Nepal and Bhutan, has had anything other than minimal economic links with the giant neighbour, and a period of almost continuous regional tensions. However, many of those tensions – the dispute over Kashmir or the civil war in northern Sri Lanka, for example – have involved a conflict of perceived interests between India and one or more of its neighbours.

It may seem absurd to talk of countries like Pakistan and Bangladesh, each with populations rising above 110 million, as

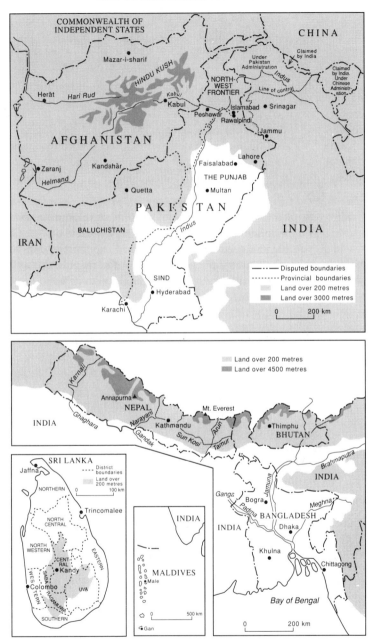

*Figure 3.1*  The peripheral states of Asia

'peripheral'. Not only do they have their own political sovereignty, but also they are sufficiently large to contain in themselves great regional diversity, their own regions of growth and stagnation, and their own 'cores' and 'peripheries'. Indeed, to group them together with other countries of the periphery such as Bhutan or the Maldives, with populations of 1 million and 210,000, and occupying territory as varied as the foothills of the Himalayas and the coral atolls of the Indian Ocean respectively, may suggest a grouping of convenience rather than of analytical validity. Yet at different scales each country and each of their internal regions is locked into a wider system of economic and political interchange. As B.H. Farmer wrote of Nepal

> adherents of the dependency school of underdevelopment see it ... as a periphery of a periphery, a dependency of India which is in turn a dependency of the world capitalist system; while within the country the valley around Kathmandu acts as the core within a periphery of a periphery to even more peripheral and remote rural areas.
>
> (Farmer 1983)

Certainly there are many specific aspects which point up the differences between the individual countries surrounding India other than their population size alone. The numerically dominant Muslim states of Pakistan and Bangladesh flank India to west and east, while the tiny Islamic island state of the Maldives lies off India's south-western coast. In social and cultural terms they could scarcely be more different from predominantly Buddhist Sri Lanka and Bhutan, or the only Hindu kingdom in the world, Nepal. To the west, Afghanistan has continued to act as a frontier zone between South Asia and West Asia. Of marginal economic signifi-cance to the rest of South Asia, and with its own unique social and political characteristics, it has been a region of vital strategic significance to the rest of the sub-continent throughout the period since 1965.

Despite the contrasts between the individual states, in practice it is impossible to understand some of the major changes in the geography of the countries of the periphery since 1965 without reference to their relationships with India. Though the specific nature of those relationships has to be understood in their own regional and historic bilateral context, they have combined to give a remarkably strong regional framework throughout the period. In

1979 President Zia Rahman of Bangladesh made the first moves towards setting up the South Asian Association for Regional Co-operation (Muni and Muni 1984). He was anxious to convert the links binding each of the countries to India into a framework for peaceful co-operation rather than of destructive competition. Today that hope is still a long way from realization (Bradnock 1990).

## POPULATION AND DEMOGRAPHIC REGIMES

Despite their diverse cultures and economies, all the countries of the periphery have experienced remarkable population growth since 1965. Death rates in some regions were still over twenty-five per thousand in 1965, although they were already significantly lower in some of the major urban areas. Today they have fallen to under ten per thousand in most cities, and while in remote rural areas of Nepal, Bangladesh or Bhutan they remain above twenty per thousand, the overall decline has reflected continuing improvements in diet and health care. Experience elsewhere in the world would suggest that at this early stage of the demographic transition birth rates would not have declined as rapidly. That has proved to be the case. Only in Sri Lanka have birth rates fallen generally to twenty-five per thousand or less, but as death rates had already fallen to six per thousand in 1980 overall population increase remained high. Elsewhere both birth rates and death rates remain much higher. Estimates in Pakistan for example suggest that in many rural areas birth rates are over forty per thousand and show little sign of falling.

Governments in several of the countries have adopted family policy programmes. Former President Ershad's government in Bangladesh, for example, adopted an education programme for Muslim leaders in the hope of persuading them to teach villagers that family planning was not against the faith of Islam. But despite official programmes overall rates of population increase remain close to 3 per cent. If that rate of growth continues for the foreseeable future the implications are striking. Actual population growth is shown in Table 3.1, with World Bank predictions for the year 2000. The final column shows the predicted size of the total population that will be reached before increases level off (World Bank 1991).

The absolute figures that result from those changes look daunt-ing, and the problems related to the growing demand for limited land resources have become increasingly obvious. Both Pakistan

*Table 3.1* Population growth in South Asia 1965–2000

| Country | Growth 1965–80 (% per annum) | Growth 1980–6 (% per annum) | Growth 1986–2000 (% per annum) | Population (mn) 1986 | Population (mn) 1990 | Population (mn) 2000 | Size of stationary population |
|---|---|---|---|---|---|---|---|
| Afghanistan | 2.4 | n.a. | n.a. | 15.5 | n.a. | n.a. | n.a. |
| Bangladesh | 2.7 | 2.6 | 2.5 | 103 | 114 | 145 | 342 |
| Bhutan | 1.6 | 2.0 | 2.2 | 1 | 1 | 2 | 4 |
| Nepal | 2.4 | 2.6 | 2.5 | 17 | 19 | 24 | 63 |
| Pakistan | 3.1 | 3.1 | 3.0 | 99 | 113 | 150 | 423 |
| Sri Lanka | 1.8 | 1.5 | 1.5 | 16 | 17 | 20 | 30 |
| India | 2.3 | 2.2 | 1.8 | 781 | 846 | 1002 | 1695 |

*Source:* World Bank (1991).

and Bangladesh can expect populations of well over 200 million by the year 2021 if there is no slow down in growth rates. As Cassen (1978) has shown, such growth implies high costs at the national level in terms of demands on investment to achieve even minimal improvements in living standards. However, these figures need to be interpreted in the widely differing contexts of the specific countries of the region. The physical environments, for example, vary from the deserts of Pakistan to the deltaic landscape of Bangladesh. Population densities vary enormously. In Baluchistan, which itself covers 350,000 sq km, an area equal to that of Germany, the population is just over 6 million. Bangladesh, on the other hand, has 110 million people in an area the size of England and Wales. Nepal, whose agricultural land has been pushed further and further up the mountainous slopes of the Himalayas, faces the problem of trying to increase agricultural output in one of the most unstable physical environments in the world. With the Himalayan ranges still rising by up to 6 cm a year (Ives and Messerli 1989), rock slides, the larger ones often precipitated by earthquakes, contribute to a naturally high rate of erosion and denudation. In the wholly contrasted environment of the Maldivian atolls population growth is also pressing on the limited land area of the tiny islands, which barely break the surface of the sea and which some authorities suggest are threatened by even a slight rise in sea level.

In many regions of the periphery, population growth has served to highlight the cultural and ethnic diversity of the modern states. In Baluchistan, the North-west Frontier and the Northern Areas of Pakistan, and in the Chittagong Hills of Bangladesh, tribal groups have been under increasing pressure to adapt their traditional ways of life in the face both of political and economic change (Johnson 1982). Since 1979 Pakistan has been home to over 3 million political refugees from Afghanistan, and by the early 1990s there was still no sign of their return. In Chittagong, settlement of agriculturalists from the Bangladesh plains was increasingly encroaching on lands which had been restricted to tribal use. The difficulties faced by the tribals has been increased by the building of the Kaptai Dam, completed in 1965. The benefits of this scheme have been electricity generation for use as far away as Dhaka and the effective elimination of floods in the previously flood-prone Karnaphuli valley. However, some of the costs have been borne by the Chakma tribes, much of whose essential valley-bottom land has been flooded, leaving them only the uplands of the watersheds (Anti Slavery Society 1984).

Several countries of the periphery have been severely affected by political disruption, sometimes bordering on civil war. By far the worst affected has been Afghanistan. The coups and counter-coups in the 1970s all reflected the interleaved influences of internal political faction fighting followed rapidly by external involvement. The Soviet invasion of December 1979 brought the United States and Pakistan into direct involvement with Afghanistan's civil war. It continued even after the final withdrawal of Soviet troops in 1989. As one result of the invasion over 3 million Afghan refugees are in Pakistan and over 1 million more in Iran. Yet the total Afghan population is probably increasing none the less, to an estimated present figure of 18 million. The shortage of land available for cultivation and the lack of investment for well over a decade have contributed to Afghanistan's dependence on two basic sources of support: Soviet aid and a flourishing black market. The most important element of this is the trade in opium, which American sources put at 750 tonnes in 1989, the world's second largest after Burma (*Asia Yearbook* 1990).

Competition for land as a result of rapid increases in population has played a part in the deepening unrest between Sinhalese and Tamils in Sri Lanka. In 1965 Sri Lanka had just embarked on a scheme for colonizing its Dry Zone. At that point its total population was nearly 11 million. In medieval Sri Lanka much of the north and east of the island had been settled and cultivated far more intensively than it was by the mid-twentieth century. Wars, probably accompanied by a rapid increase in the prevalence of malaria, had caused much of that land to become largely depopulated. During the 1950s the government embarked on comparatively small schemes such as the Gal Oya to irrigate and recolonize waste land. The Sri Lankan population continued to grow through the 1970s and 1980s, and the lack of alternative employment outside agriculture encouraged the government to go ahead with the massive Mahaweli Ganga development project. Although this multipurpose scheme was designed to provide both power and irrigation, as it made possible the resettlement of land in the north and east of the island, it enabled the government to settle Sinhalese into lands that the Tamil community regarded as traditionally its own, changing the ethnic balance and providing one more element in an already worsening relationship between the island's main communities. By the late 1980s when the scheme had been in operation for over a decade it had settled fewer than

12,000 farmers or 70,000 people, yet during the same period the total population of the island had grown by more than 6 million (Swan 1987).

Many features of the population structure of South Asia's periphery illustrate the strength of the ties of each of the countries to India. One of the legacies of the Partition which created Pakistan and India in 1947, and ultimately gave shape to Bangladesh, was a sub-continent whose political divisions cut through pre-existing social, cultural and economic links. This was particularly evident in Bangladesh, where population growth in the north-eastern districts without a corresponding increase in opportunities for productive work, had encouraged emigration, notably into the Brahmaputra Valley of Assam, ever since the late nineteenth century. The creation of Pakistan and the official closing of the border failed to stop the movement. In the early 1980s the scale of migration of Muslim settlers from Bangladesh became the focus of massive political unrest in Assam, an issue which remains unresolved but which was one contributing factor to the distrust which marked Indo-Bangladesh relations throughout the decade.

All of Sri Lanka's people can trace their origins back to India, the Sinhalese to the north of India and the Tamils to the south (de Silva 1981). Today the Sinhalese make up 74 per cent of the total population. Sri Lanka's Tamil population comprises the long-settled Tamils of the north and east (12.6 per cent) and the recent migrant workers on the tea plantations in the central highlands (5.5 per cent). In 1965 the then Indian and Sri Lankan Prime Ministers, Lal Bahadur Shastri and Mrs Sirimavo Bandaranaike, signed a repatriation agreement under which nearly half a million tea plantation workers and their families would be returned to India, the remainder being granted Sri Lankan citizenship. By the early 1990s over 350,000 adults from this Tamil community had been repatriated to India, but despite the high volume of repatriation, natural increase ensured that there were as many Tamils on the tea estates in 1991 as there had been twenty-five years earlier. Sri Lanka's ethnic composition was further complicated by the presence of the so-called 'Moors', Tamil-speaking Muslims of Indian-Arab descent, who came as traders to Sri Lanka's east coast, and who now number over 1.1 million (7.7 per cent). The problems faced by such minorities are common to all the peripheral states (Jayawardene 1987).

ROBERT W. BRADNOCK

# URBANIZATION AND ECONOMIC CHANGE

If political change and the rapid growth of population have often caught the headlines during the last twenty-five years, there have been less reported but none the less important economic changes across the region. Population growth in itself gives only the barest outline of the range of demographic changes which have taken place in each of the countries. Despite the fact that all the countries of South Asia remain predominantly rural, the urban population has been growing rapidly, with far-reaching social and economic implications. In 1965 some countries in the periphery – Bangladesh, Nepal, Bhutan and the Maldives – had under 10 per cent of their population living in towns and cities. Even Pakistan had less than 20 per cent classified as urban.

Today urban growth has changed the balance between town and country and is bringing with it rapid social change. The most urbanized regions – in Pakistan and Sri Lanka – now have over one-third of the population living in cities. The largest cities – Karachi, Lahore, Dhaka – have seen tremendously rapid growth. In 1965 Dhaka was estimated to have a population of fewer than 500,000. By 1981 the figure had risen to over 3.4 million and by 1990 it was estimated to have reached 4.9 million (Bangladeh Bureau of Statistics 1991). Karachi, with a population of about 10 million, is by far the largest city in the peripheral states.

Much of the urban growth in towns right across South Asia has taken place in an unplanned, spontaneous way, a result of the combination of migration, natural increase and the reclassification of previously rural places as urban. Rural to urban migration has contributed more than a half of the increase in some areas. Partition between India and Pakistan in 1947 led to a flood of migrants into the towns of both West and East Pakistan, though by far the larger number went to the cities of Sind and Punjab in the west. Those refugees – the 'muhajirs' – brought a new foreign element to the cities of Pakistan, changing their social character and creating a source of competition for space and for work which has fostered divisive tensions in Pakistan's urban communities. At the same time some Muslims from Bihar moved to East Pakistan, but when East Pakistan seceded to gain its independence as Bangladesh in 1971 the Bihari community was left stranded and isolated as a despised community of outsiders in an overwhelmingly Bengali state.

Political tension is just one dimension of urban change which has characterized the period since 1965. In part the rapid growth of towns has reflected the search for work. In Pakistan such work came at first from the increasing development of small-scale industries servicing the irrigated agricultural tracts of the Punjab and Sind. Such opportunities have continued to grow, and Green Revolution changes in agriculture have set up a steeply increasing demand for items like electric and diesel pumpsets, bicycles, and mini-tractors. Centres like Lahore, which had long been focuses of craft and other small-scale industry, capitalized on the new opportunities, just as their neighbours in Indian Punjab and Haryana diversified their industrial base (Johnson 1979).

Population growth and rapid economic change brought new housing and transport facilities to many of the larger towns in the region. Yet another common feature of urban growth throughout South Asia during this period was the sprawling development of slums and squatter settlements. The peripheral countries were no exception. Some estimates suggest that as much as 50 per cent of Dhaka's population lives in 'informal' housing or squatter settlements. Facilities in all such settlements – in Dhaka, in Karachi or Lahore, in Colombo – are totally inadequate. Water supply is normally limited and contaminated, sanitation is non-existent and the transport system stretched far beyond capacity. Yet alongside the sprawling bustees and katchi abadis (names for shanty slums) there has been a remarkable growth of modern housing and offices. The centres of Karachi, Dhaka or Colombo now have the high-rise office blocks typical of any large commercial city. In Pakistan the comparatively liberal import policy of successive governments has allowed roads to be filled with modern, foreign – largely Japanese – cars and motorbikes, adding to the impression of rapidly modernizing cities.

The scale of urban population growth has often transformed the pre-existing city. Although it is the home of one of the world's oldest urban civilizations in Moenjo Daro and Harappa, most of Pakistan's urban centres owe their origin to the spread of canal irrigation in the nineteenth century. Some of those market towns have been transformed. Faisalabad, for example, now a city with 1 million people, has become a major centre of the textile industry. Town planning has ensured that in cities such as Lahore or Pakistan's new capital of Islamabad, formal housing colonies are spaciously laid out. Like Chandigarh in India, and parts of Dhaka in Bangladesh, Islamabad was designed by foreign architects to distinctly non-South Asian specifications. The inflow of capital from migrant workers in the Gulf saw remittances to Pakistan, Bangladesh and Sri Lanka provide a significant new source of income for hundreds of thousands of people. Table 3.2 demonstrates that the remittances made an important contribution to both national and local economies, especially in Pakistan.

*Table 3.2* Remittances from the Middle East

| Date | Bangladesh ($ mn) | Sri Lanka (Rs mn/% GDP) | Pakistan ($ mn/% GDP) |
|------|------|------|------|
| 1977 | 11 | 109 (0.6%) | 435 (2.9%) |
| 1980 | 159 | 2,518 (3.9%) | 1,667 (5.9%) |
| 1985 | 363 | 7,920 (5.2%) | 2,069 (6.7%) |

*Source:* Amjad (1989).

The Gulf War demonstrated that benefits of such remittances could be offset by the increased dependence of the South Asian periphery on events completely beyond their control. Ziring (1991) suggested that the Gulf crisis presented Pakistan with 'a monumental crisis', with a shortfall of between $1.5 billion and $2 billion in foreign exchange resulting both from a drop in remittances and a decline in exports to Iraq. Tens of thousands of Pakistanis returned home. The position for Sri Lanka was even worse, with more than a quarter of its tea exports normally going to Iraq and 100,000 domestic servants and manual labourers in Kuwait forced to flee (Singer 1991).

The cash from remittances has often been invested in housing, converting the fields on the suburban fringes of some of the largest cities to great areas of residential housing, often with new community facilities. Yet in the interstices – along roads, railway lines, and any patches of open government land – landless labourers from rural areas pack into the cities in search of work.

The growth of Pakistan's cities, their modernization and industrial development, conceals the extent to which they are also centres of deep social and political division. In the southern province of Sind no government of Pakistan has been able to weld together the immigrant Muhajir population with the local Sindis. The migrant community from the northern plains of India has retained its cultural and economic distinctiveness since 1947. In the last twenty-five years modernization of the urban economy, far from reducing social divisions, has heightened tensions between communities.

The speed of urban growth has not always been directly related to industrial expansion. The pace of industrialization has varied widely both between countries and within them. In 1965 only Pakistan had an industrial base. It was also the only country which had developed significant mineral resources, natural gas already making an important contribution to meeting both direct energy needs and as a feedstock for Pakistan's fertilizer industry. Johnson (1979) has suggested that 'land, water and a warm climate are Pakistan's main natural resources'. Since the mid-1960s gas and oil have been developed, but other mineral resources are limited; limestone is quite widespread and rock salt is common in the Great Salt Range, providing potential for a major chemical industry. There are workable deposits of copper, but only very poor resources of iron ore and coal.

Natural gas is by far the most important single resource, vital for meeting Pakistan's growing energy needs as well as providing a feedstock for the fertilizer industry. Three-quarters of the reserves are in the Indus Plains, the largest single gas field being at Sui near Sibi. The developing pipeline network has enabled gas to be piped to both Lahore and Karachi. While extensive oil exploration has been carried out production is still limited, meeting about 20 per cent of the country's needs by 1990. Mineral energy sources are in short supply throughout the peripheral countries. Bangladesh also has natural gas and may have more gas and oil off-shore in the Bay of Bengal, but Nepal, Bhutan, Sri Lanka and the Maldives have virtually no exploitable mineral energy sources. Given those severe limitations, and the cost of importing fuel, major efforts have been made since the mid-1960s to develop hydro-electric potential (HEP).

By far the most significant developments have taken place in Pakistan, which has been estimated to have about 10,000 MW of hydro-electric generating potential. The chief stimulus to development came with the signing of the Indus Waters Treaty by India and Pakistan in 1960. In view of the bitter political distrust between the two countries the Agreement was a remarkable achievement. Under the plan India was to be allocated the waters of the three eastern tributaries of the Indus, the Beas, Ravi and Sutlej, which through the nineteenth and early twentieth century had been developed to irrigate western Punjab. This loss was to be made good by developing the western tributaries, the Chenab and the Jhelum, and the Indus itself, the cost being borne by massive investment from the World Bank. The scheme involved building some of the world's largest dams – at Bhakra Nangal in India, and at Mangla and Tarbela in Pakistan, as well as hundreds of kilometres of major link canals. In addition to its contribution to increasing the irrigated area in Pakistan, the completion of the Tarbela Dam on the Indus added 2,000 MW to Pakistan's generating capacity, over one-third of the total potential now having been developed. The Mangla Dam on the Jhelum produces 800 MW. However, all the dams in the Himalayan foothills suffer from the very high levels of erosion in the mountains and are filling rapidly with silt. The life expectancy of the Mangla Dam is now less than sixty years. Furthermore, the area is at constant risk of earthquakes, which adds greatly to the cost and complexity of further development.

HEP development in all the other countries of the region faces similarly complex problems, though their specific nature varies from country to country. In Nepal there is enormous potential in the deeply cut valleys, but the population is so scattered that transmission costs are extremely high. However, as with Bhutan, Nepal has the potential of exporting electricity to the large market of India, where despite steady increases in generation demand for electricity continues to outstrip supply in nearly every state. Experimentation with small-scale schemes in Nepal has had the added incentive of allowing development with relatively low capital costs. The larger schemes in Pakistan, Bangladesh and Sri Lanka have been possible only with massive investment of foreign aid: from the World Bank for the Mangla and Tarbela Dams, from Japan for the Kaptai Dam in Bangladesh, and from a consortium including British aid for the Victoria Dam project in Sri Lanka. In addition to their very high capital costs, however, such dams are increasingly seen as posing severe environmental problems.

Energy resources, especially of gas, have been the basis of some heavy industrial expansion, especially in fertilizer production. Again, Pakistan and Bangladesh have been best placed to exploit the opportunity. Multinational investment in the 1960s from major oil companies has been followed more recently by investment from Saudi Arabia and the Gulf. Pak-Saudi Fertilizers at Mirpur-Mathelo in Sind is one example, but there are several other big plants, such as those at Mianwali, Faisalabad, Multan and Sukkur. The national oil refinery at Karachi has a capacity of nearly 5 million tonnes, and there is a widening range of medium-scale industrial projects. However, Pakistan has followed a quite different industrial development path from its neighbour India, with much less emphasis on local production of the full range of industrial products and a more open door policy to imports, especially of goods such as cars, motorbikes and electrical goods. There has also been a very large measure of illegal duty free import through the North-west Frontier region, and it is possible to get a wide variety of imported goods. Traffic on Pakistan's roads is also quite different from that in India, Japanese imports dominating both goods and passenger traffic.

In the early years after Independence successive governments tried to encourage private enterprise by protecting 'infant' industries and giving considerable help throughout the Pakistan Industrial Development Corporation, which made capital for investment available. Much of the investment that took place remained in the hands of Pakistan's wealthiest twenty-two families. In the early 1970s Zulfikar Ali Bhutto introduced widespread nationalization, a policy reversed once more under the regime of President Zia ul Haq (Burki 1988). Over half a million workers depend on manufacturing industry in Pakistan today. Karachi dominates industrial activity in the south, with Hyderabad a comparatively minor centre, and Lahore the north, Faisalabad and Multan being important secondary centres. Nearly 40 per cent of industrial jobs are in the textile industry, Faisalabad and Multan leading the way. Faisalabad is an important milling centre while Multan specializes in cotton ginning. Karachi has a diverse industrial base, food, metal and engineering industries all being important. The recent completion of an integrated iron and steel mill at Pipri, 40 km east of Karachi, built with Soviet assistance, adds to Sind's industrial importance. Lahore has iron works, textiles, chemicals, printing, food and footwear industries.

Processing of raw materials appeared to offer the best hope of achieving some measure of industrial development in the other countries of the periphery. Bengal's lucrative cotton industry, destroyed by colonial trading policies and Maratha disruption of traditional internal markets, had disappeared long before Independence in 1947. By then raw jute had long been exported from Bengal – first to Scotland, then to Calcutta. After the creation of Pakistan new mills began to be set up, and expansion continued after

Bangladesh gained its independence in 1971. There have been some opportunities for diversifying. Ready-made garments – what would once have been termed piece-goods – now constitute more than a third of Bangladesh's exports, utlizing some of the the abundant cheap labour. Some attempts are also being made to encourage small-scale engineering industries. And the export of shrimps (about 10 per cent of all exports) now exceeds the value of raw jute. In Sri Lanka manufacturing accounted for less than 5 per cent of the Gross Domestic Product at Independence. By 1990 a number of new industries had been developed – cement, mineral sands, ceramics, cloth. These were all planned originally in the state controlled sector. The socialist government under Mrs Bandaranaike envisaged public ownership of all major industries, but the United National Party government elected under President Jayawardene's leadership in 1977 reversed this policy, moving towards a free trade economy. Among the leading sectors of the new policy was tourism, with particular efforts to exploit the superb beaches and equable climate. This programme has been severely hit by the political troubles which have dogged the island since 1983 and in 1989 tourism ceased almost completely before picking up again in 1990. In 1991 the total value of industrial production was just under Rs 50 billion, the food and beverages industry, textiles and leather, and chemicals each contributing about Rs 14 billion. Nepal and the Maldives also tried to capitalize on their resources of landscape and sunshine. The beaches of the Maldives, and the Himalayan mountains in Nepal, became the focus of a major campaign to win tourists through the 1970s.

## AGRICULTURAL CHANGE

Contrasts in agricultural change reflect the enormous contrasts in the productive agricultural potential of the physical environment. They are matched by equally great contrasts in farming techniques, crops and cropping patterns which have played an important role in the regional patterns of agricultural change which have taken pace since 1965. Wheat dominates the agriculture of the Punjab plains but is also important in Sind and North-west Frontier Province (NWFP), accounting for more than half the 12 million hectares of cultivated land. It is the staple crop, vital to a diet based on chapattis, nan and parathas. Since Independence, rice has been of increasing significance, especially as an export crop. High-quality rice such as basmati has become a high-value earner of foreign exchange, and rice is particularly well suited to the often salty soils of the lower Indus plains. This is slightly ironic, in that new rice varieties thriving on irrigation and sunshine have been able to increase yields more in the non-traditional rice areas. The 'IRRI-Pak' varieties are derived from the varieties pioneered by the International Rice Research Institute in the Philippines which was most anxious to increase rice in the cloudier and wetter traditional rice-growing areas of Asia. Mexi-Pak wheat and IRRI-Pak rice now account for almost the entire cropped irrigated area. The effect on production has been dramatic. In parts of Sind wheat yields had reached over 2,100 kg per ha in 1987 and the total output reached nearly 13 million tonnes, while rice production stood at over 3.5 million tonnes. On the unirrigated land of the valleys (known as *baranni* land) in NWFP and Baluchistan local varieties of wheat and hardy crops such as barley, maize, millet and gram (a pulse) predominate. These have not seen the same dramatic yield increases.

Pakistan, and especially the Punjab, rapidly became the focus of Green Revolution technology. However, Sri Lanka also experienced a rapid adoption of new seed varieties from the late 1960s. It is striking that Sri Lanka had not produced enough food to meet the needs of its population since the eighteenth century, yet in many respects it had been the most obviously prosperous state in South Asia. After Independence the growing food deficit was an enormous economic problem, imports of foodgrains accounting in the 1970s for half the foreign exchange earning. Attempts to increase rice production ranged from wide-ranging land reform to the introduction of high yielding varieties of seed. By 1970 three-quarters of all rice sown was high yielding and by the early 1980s there was virtually a 100 per cent take-up of new varieties. Yields have increased to over 3.5 tonnes per ha and production showed a marked increase, rising towards 80 per cent of domestic needs despite the speed of population growth.

Much of Sri Lanka's relative prosperity had come from the cash crops of tea, rubber and coconuts, which continued to earn the lion's share of Sri Lanka's foreign exchange earnings from Independence right up to the present day. In 1950 they accounted for 96 per cent of total exports. By 1978 they still contributed 73 per cent, and by 1991 over 50 per cent of foreign exchange earnings came from these three products alone. Tea has suffered from inadequate investment and fierce competition from expanding production in other countries of lower price, lower quality tea. The area cropped under tea fell from 244,000 ha in 1979 to 221,000 ha in 1987. Production rose sharply from 140 million kg in 1948 to 230 million kg in 1965 but then declined to around 180 million kg in 1983. Since then it has risen again to over 210 million kg in 1991. Tea alone still accounts for Rs 10.6 billion of exports, a third of the total value of exports, followed by rubber (Rs 3 billion) and coconuts (Rs 1.4 billion).

The greatest test for agricultural development in the periphery remains Bangladesh. For centuries Bangladesh was regarded as one of the most fertile regions in South Asia. Annual inundations by the floodwaters of the Ganga, Brahmaputra and the Meghna rivers helped soils to renew their fertility, and traditional agricultural technology developed highly sensitive adaptations to even the slightest variations in environmental conditions. Double and some-times triple cropping were practised where water was available, though the five month dry season restricted winter cropping to under 1 per cent of the total acreage. By the time it became independent in 1971 Bangladesh was having to import at least 30 per cent of its food supplies, at a time when the world demand for jute, the only significant export it had with which to pay for those imports, was contracting. Although improvement was far from rapid, the late 1980s have seen some increase in output. Cata-strophic flooding recurred through the 1980s. In 1988, for example, one-third of the country was under water. Yet measures to improve winter cropping and modest success with high yielding varieties and fertilizer adoption programmes have lifted production close to self-sufficiency, despite the rise in population. Diets remain extraordinarily limited and malnutrition levels high.

The most remarkable feature of the development process in the periphery since 1965 has been the fact that any significant improve-ment should have taken place at all. With endemic political insta-bility, massive dependence on external trade in products for which the terms of trade moved consistently against their interests, and severely limited internal resources for investment, many regions have shown a capacity to achieve economic development in spite of the many obstacles they face. However, that development has been both spatially and socially uneven. Not only has land distribution remained unequal but also landlessness has increased throughout the region. In Bangladesh the average size of land holdings has decreased to well under 1 ha per person, each farm being divided into up to twenty fragments. In Pakistan, on the other hand, large landlords continue to dominate agriculture, the semi-feudal structure of landownership, notably in Sind, having survived President Bhutto's land reforms of the 1970s without fundamental change (Burki 1988).

## PRESSURE ON THE ENVIRONMENT

The varied and fluctuating performance of the periphery in terms of economic development has left larger numbers of their people living below the poverty line, however arbitrary the definition, than twenty-five years ago. The pressure to improve production both in agriculture and industry has therefore been enormous. However, where development has taken place it has often produced its own problems. The most striking example of these have been the depletion of agricultural resources that have taken place in large areas of Pakistan's irrigated agriculture. Canal irrigation, introduced from the mid-nineteenth century in the region around Lahore and spreading south and west through a series of increasingly large schemes, made possible the settlement and cultivation of vast tracts of land which had formerly been scrub jungle. By the mid-1960s, however, some of those areas had been suffering the twin problems of salinization and waterlogging for over fifty years.

Many of the early schemes in India and what is now Pakistan suffered from poor drainage. The new canals sometimes cut across natural lines of drainage, damming up the flow of surface water. The irrigation itself was carried out simply by flooding the fields from canals, most of which were unlined. Some estimates suggested that as much as a third of the water that left the barrages and dams was lost by seepage through the unlined canal beds, and a further third by evaporation. During the irrigation of the fields themselves more water also percolated down to the underground water table. In some areas this rose rapidly, causing fields to be waterlogged. The very high daytime temperatures and low humidity give extremely high evaporation rates over much of Pakistan through most of the year, and thus the stagnant water in the waterlogged fields was often being dried out simply by evaporation, in the process leaving behind the minerals and salts which had been in solution. It rapidly became apparent that huge damage was being done to some of the most productive irrigated areas. B.L.C. Johnson quotes one engineer as saying that 'heroic measures are essential if the Punjab is not to be destroyed' (Johnson 1979).

Before Independence a number of measures were introduced to tackle the problem, including planting deep-rooted species of trees such as eucalyptus along canal banks to lower the water table, and restricting the use of canals. However, the scale of the problem was still enormous. Estimates suggest that in the mid-1970s over a third of Punjab and Sind's irrigated land was moderately to severely affected by waterlogging. While Punjab was less severely affected by salinity, virtually all of Sind was affected to some degree. After 1947 the Government of Pakistan set up the Water and Power Development Authority (WAPDA), which planned a series of Salinity Control and Reclamation Projects (SCARPS) in the most seriously affected areas. The first of these came into effect in 1959 between the Ravi and the Chenab rivers in Punjab. The basis of the programme has been to sink tube-wells at frequent intervals and to drain off the water through specially cut channels into the rivers, thereby lowering the water table. The cost and the effectiveness of such schemes depends in part on the salinity of the underlying ground water itself, but in any event it is important to reduce the water table so that the movement of water is from the surface downwards, washing out salt, rather than by capillary action upwards, which deposits salt at the surface as the water evaporates. In areas where the ground water proved sweet it was possible to use it for irrigation, reducing dependence on the canals and being directly productive. However, much of Sind is underlain by severely saline groundwater, which cannot be used for agriculture and must simply be drained away, greatly increasing the costs of the programme. In the mid-1970s, official surveys carried out by SCARP suggested that despite the programmes already implemented, 24 per cent of the cultivable area in Punjab and 19 per cent in Sind were still severely waterlogged, the corresponding figures for the area suffering severe salinity problems being 13 per cent and 17 per cent respectively. Although these suggested considerable improvement over earlier estimates they indicated the scale of the remaining task. The challenge is an unending one, of vital significance to Pakistan's economic survival.

Pakistan's agriculture also has to cope with the natural hazards of flooding. The highly variable flow of the major rivers is still far from being completely controlled. Furthermore, all the rivers carry enormous quantities of silt. It has been estimated that the lower reaches of the Indus carry four times as much silt as the Nile or twice as much as the Missouri, enough to cover 100 sq km to a depth of 1 metre every year or 435 million tonnes (Ives and Messerli 1989).

Although the specific nature of the environmental costs of development varies from country to country there is a growing awareness of the delicate interrelatedness of the agricultural and environmental systems. Recent work by Ives and Messerli suggests that it is simplistic to assign all the problems of environmental change to single causes such as deforestation in the Himalayas. They point out that research in the Himalayan belt illustrates the complexity of the interrelationship between traditional terraced farming and both forest cover and slope stability. In a region of continuing natural dynamic instability, there is evidence to suggest that in parts of Nepal, for example, farming has helped to stabilize slopes and reduce erosion, evidence that flies in the belief widely popular in the late 1980s that the Himalayan region was in ecological crisis. Yet at a local level removal of forest cover has often restricted access to resources vital to the poorest people, especially fuel, a lack which has not yet been compensated by the development of alternatives. Today it is probably true that the extensive margins of cultivation have been reached, adding to the need to intensify the use of land already under cultivation. The process has already made remarkable progress in Nepal. In 1965–6 1.8 million ha were cultivated with a cropping intensity of 108 per cent. By 1985–6 the cultivation covered 2.4 million ha, but the cropping intensity had risen to 166 per cent. Bangladesh also experienced increasing intensity of land use during the same period, from 137 per cent to 150 per cent, but in Bangladesh the total area cultivated actually fell slightly. Ives and Messerli conclude with reference to Nepal that 'intensification has to be achieved in all three major land use types – cropped land, grazing land, and forests. Land must be allocated to uses which are not degrading and which, at the same time, represent the best and most productive use of that land' (Ives and Messerli 1989:198). The same conclusion applies to all the peripheral states.

In Bangladesh the environmental problems are made particularly acute by the high population densities, but their causes are complex. Its coastal regions are subject to the risk of immense damage from cyclones and saline incursions up the distributaries of the Ganga-Brahmaputra system. The building of the Farakka Barrage in India to channel water down the Bhagirathi-Hooghly past Calcutta, helping to tackle Calcutta's acute water problems, has exacerbated those of the dry western region of Bangladesh. As the dry season flow down the main channel has been curtailed, so sea water has penetrated ever further inland, making previously fertile land become unusable. The challenge is made more complex by the enormous scale of riverine flooding during the monsoon period, and the difficulty even of draining away rain water, which can contribute to deep flooding even in the absence of overspill from the rivers. A variety of protective measures have been advocated. From its independence in 1971 the Bangladesh government argued that, as over 90 per cent of the watershed of the rivers which built up its delta lay in India and Nepal, the problems of the Ganga basin should be tackled on the basis of a plan agreed by all three countries. India, however, refused to join anything other than bilateral negotiations, fearful that any broadening of discussions could lead to its being outnumbered. As a result virtually no progress was made in coherent planning for the basin as a whole. In 1990 Bangladesh looked towards a massive new flood protection and irrigation scheme within its own borders, but it remained unclear whether the costs involved could ever be matched by equivalent benefits (World Bank 1989).

In the controversies surrounding the implications of global warming, politically motivated rhetoric often replaced scientific analysis. Many fundamental questions remain unanswered. The possible rise in sea level – itself questioned by some authorities – has been held to threaten the inundation of large areas of both Bangladesh and the Maldives. However, both are active natural systems, highly responsive to change. There is evidence to suggest that if sea level starts to rise, land levels may rise as fast. In the Maldives this could occur through the growth of the coral which forms the islands. In Bangladesh the delta rivers carry over 1 million tonnes of silt per day, and some argue that a rise in sea level will be accompanied by a corresponding increase in rates of deposition, building up the delta accordingly. Despite the great uncertainties, the pressures on the environment are clear in many parts of the region.

## THE CULTURAL CONTEXT

Political independence for the new nation states was widely accompanied by unspecified ambitions of 'modernization' and 'development', expressed alongside a determination to sustain local cultural traditions. One of the key symbols and programmes for modernization throughout the region has been education. Yet despite government efforts to extend education throughout the region the twenty-five years since 1965 have seen the problems accumulating faster than the solutions.

The linguistic development of Pakistan reflects the isolated and fragmented nature of the tribal populations and the clearly identifiable regions of settled agriculture on the plains. Apart from the Dravidian Brahui all Pakistan's languages belong to the Indo-European family. Urdu, the national language, developed as a synthesis of spoken Hindi and written Persian as the language of the followers of the Mughal court. However, the heartland of Urdu lay not in modern Pakistan but in north central India. It was the language of the migrants (the *mohajirs*) who came into Pakistan in 1947 at the time of Independence, but the two major languages of Pakistan were Panjabi and Sindi. They remain numerically the most important, although Urdu is the national language. Even Bangladesh, with a greater degree of linguistic unity than any other peripheral state except the Maldives, has several tribal languages alongside the dominant national language of Bangla (Bengali).

Even at the most basic level of literacy the statistics continue to illustrate the relatively small impact made on traditional patterns of life. By 1981 about 55 per cent of males and 38 per cent of females who lived in Pakistan's towns were classified as literate. In rural areas the figures dropped to 25 per cent for males and 7 per cent for females. In rural Baluchistan the figures of female literacy fell to under 2 per cent. The average figures were even lower in Bangladesh and Nepal. Sri Lanka has made by far the greatest progress, over 90 per cent of males and 83 per cent of females being classified as literate in 1980.

Literacy is only the barest of measures of cultural change and diversity, which have played a major part in the continued political spasms which have shaken every state of the periphery since 1975. Links which some of these ethnic minorities retain with India at the centre have often reinforced the problems of national identity which ethnic diversity has posed. Civil wars in Sri Lanka after 1983 between the Sinhalese and Tamils, or in Pakistan during the 1970s with the Baluchis, have been framed around demands for the preservation of ethnic identities.

# POLITICAL CHANGE IN SOUTH ASIA

Although all the countries of South Asia have a degree of autonomy, at many points their political interests intersect. Throughout the period since 1965 internal tensions within the states of the region spilled over into wider conflict. Such conflicts were sometimes deepened by the activities of the superpowers, still locked in Cold War confrontation and seeking allies and supporters for strategic purposes.

From Independence India was seen by many outside South Asia as the dominant power of the region. It was not a perception shared by many Indians themselves, and fears of weakness were deepened by their defeat at the hands of the Chinese in the Himalayan border war in 1962. Their central position in South Asia itself gave India an apparently commanding control over the land mass, and the long coastline gave potentially commanding control of the Indian Ocean. In practice India feared that the presence of the superpowers in their wider sphere of influence would add Cold War conflict to existing regional tensions. For India the reality of those tensions related not just to the agonies of Partition from Pakistan in 1947, but the continuing conflict between the Islamic ideology on which Partition was based and India's perceived need to sustain a secular state if its own diverse population were to continue to identify strongly with the Indian state. Non-alignment was one of the key means by which India saw itself remaining outside the dangers of Cold War confrontation, while allowing itself to proclaim its own version of secular socialism and peaceful coexistence as a means of securing a new international order.

The quarter century started with violent conflict between Pakistan and India. Kashmir, whose accession to India had remained bitterly disputed by Pakistan ever since Independence, had been drawn increasingly strongly into the Indian Union. The death of Indian Prime Minister Nehru encouraged the Pakistan President Ayub Khan and his military government to launch a war of 'liberation' in Kashmir in the autumn of 1965. Far from providing the military government with a quick victory it ended in stalemate on the ground. The Tashkent Agreement, negotiated through the mediation of the Soviet Union, brought an end to the fighting. However, it deeply damaged President Ayub Khan's credibility at home and contributed to the growing internal political crisis in Pakistan.

None of the countries of the periphery other than Sri Lanka

has had sustained periods of democratic government since 1965. President Ayub Khan's military government collapsed in 1969, but the return to democracy took place only after Bangladesh had succeeded in gaining its independence. In the intervening period, the Pakistan Army had been used to try and put down the Bangladeshi independence movement by force. Between March and December 1971 up to 10 million refugees fled from East Pakistan into neighbouring India. Using the plight of the refugees to gain world support, Mrs Gandhi fostered the Bangladeshi independence movement, giving it a home and training its *mukti bahini*, or freedom fighters. The two-week war of December 1971 saw the Pakistan army in the east routed and a government in the independent state of Bangladesh take power.

The war produced a dramatic change in the balance of power between the centre and the periphery. When Pakistan's new, democratically elected Prime Minister Zulfikar Ali Bhutto, signed the Simla Agreement with Mrs Gandhi in 1972, it seemed that the basis had been laid for a new working relationship between India and Pakistan. However, the lack of strong sense of national identity continued to plague all the states of the periphery. Within two years Mr Bhutto launched a fierce military campaign against Baluchi separatists in the south-west of the country, a civil war which lasted until the late 1970s. The occupation of Afghanistan by the Soviet Union in December 1979 produced its own flood of 3 million Pathan refugees, threatening the already difficult relationship between the North-west Frontier region and the plains of Punjab and Sind, where 80 per cent of Pakistan's population lived. Furthermore, there was no sign that the community of earlier refugees who had fled from India at the time of Partition in 1947 were being successfully absorbed into their new urban communities in Karachi, Hyderabad and elsewhere in Sind Province. Resentment against Mr Bhutto's violent abuse of power towards the end of his government were highlighted during the 1977 election campaign, making possible a subdued acceptance of his dismissal by the Army Chief of Staff, General Zia ul Haq. It was the start of a further extended period of non-democratic military rule, continuing up to his assassination in July 1988.

Violent conflict in the region was not restricted to Pakistan, and in each case India either has been directly involved or has been drawn in. Sri Lanka's Sinhalese-dominated democratically elected government had introduced a succession of measures which they saw as restoring to the Buddhist Sinhalese their rightful place in the government of an island in which the Tamil Hindu minority had an economic importance well beyond their numerical strength. The 1956 election had been fought largely on the issue of making Sinhala the sole national language. 'The ensuing period of tension culminated in the serious communal disturbances of 1958' (Farmer 1983). That was just a prelude to the slide into political chaos and civil war which occurred from 1977. Fear of the Sri Lanka Tamil community that they were suffering increasing discrimination encouraged the growth in support for extreme separatist movements. A pogrom against the Tamils of Colombo in July 1983 saw a campaign culminate in the government's determination to find a military solution to the problem of Tamil political violence. In the process, as elsewhere in South Asia, many people fled their homes, as many as 160,000 Tamils seeking refuge across the Palk Straits in Tamil Nadu, India.

The Sri Lankan government's campaign against the Tamils of Jaffna and the north-east of the island forced itself on to the Indian government's agenda, for the scale of repression of a community with strong ethnic links to one of its own ethnic communities could not be ignored. Yet the Indian government saw its interest not as supporting the separatist movement in Sri Lanka but in pacifying it, for it saw its own diversity as an open door to potentially secessionist movements. When the Sri Lankan army appeared to be on the point of a genocidal push against Jaffna in July 1987 Rajiv Gandhi's government succeeded in pressuring President Jayawardene to accept Indian troops to monitor a cease-fire, to disarm the Tamil separatists and to oversee a political solution. Finally withdrawn three years later by a new Indian government, the Indian Peace Keeping Force left behind a fragile peace and no evidence of a long-term solution.

Not one state of the periphery was untouched by political unrest. In Nepal, experiments with democracy included a party-based election in 1957, but the Congress government was overthrown in 1960 in a coup inspired by the King. His party-less *panchayati* (local village council) system of government survived until 1990, when a sudden upsurge in demonstrations forced the King to accept a return to a much fuller democratic government, culminating in the elections of March 1991. To a far greater extent than the other peripheral states, Nepal, landlocked and sandwiched between China and India, was dependent on Indian trade and Indian goodwill for its continued functioning. With up to 3 million Nepalis living and working in India, it was faced throughout the period with the struggle to achieve an extraordinarily delicate balance between asserting its rights to sovereign independence and accepting the realities of its geopolitical position.

## THE INTERNATIONAL CONTEXT

In Sri Lanka and Bangladesh cash crops sold overseas have been vital to the economy since long before they obtained independence. Tea, coconut and rubber in Sri Lanka, jute and tea in Bangladesh, had contributed 90 per cent of their foreign exchange earnings. The war with Pakistan wreaked havoc on the cash crop economy in 1971–2, reducing tea output from around 65 million lb before the war to 26 million lb. Production soon picked up again however, and by the late 1980s had reached 100 million lb. Jute production fell from well over 1 million tonnes before the war to under 800 thousand tonnes during it. However, strong international competition from artificial fibres made recovery slower and more difficult, and throughout the 1980s production hovered under 900 million tonnes. Lacking cash crops or large mineral reserves, the other countries of the periphery had even greater problems earning foreign exchange with which to pay for needed investment. The gap was made up by a combination of foreign aid and remittances from workers overseas.

Aid played a role in all the countries of the region. Throughout the 1970s and 1980s Bangladesh's entire development budget came from external assistance. In the early 1980s, for example, Bangladesh was running a current account deficit of between Tk 5 billion and Tk 19 billion a year. In 1977–8 grants totalled just over Tk 6 billion with loans contributing a further Tk 5.9 billion. Official Development Assistance totalled $1.45 billion in 1986, or $14 per capita per annum. (In per capita terms Sri Lanka was by far the largest recipient of aid in the region, with $35 per annum.) Remittances from Bangladeshis working overseas were still at the relatively low level of Tk 1.7 billion. That contribution rose sharply in the next five years to Tk 18.6 billion in 1983, with grants increasing to Tk 18.6 billion and loans (net of repayment) to Tk 11 billion. Investment was made possible only by these large-scale drafting of funds from outside. Aid in cash was supplemented by aid in kind. In 1974–5 Bangladesh received over 2 million tonnes of food aid, falling to nearly half that figure by 1986. Remittances from the Gulf countries played an important part for all the Muslim countries of the region, and for Sri Lanka, where many of the minority Muslim population obtained work outside the island (Amjad 1989).

There is no simple foreign policy stance that results from the flows of either aid or trade. By far the most important factor influencing foreign policy has been the perceived need to protect independence threatened from external forces. Pakistan, by far the largest of the peripheral countries, has always seen that threat coming largely from India. As a result it has actively sought alliances which could supply it with necessary arms and military support. To strengthen its position as a Muslim state it has tried to develop strong links with Iran and Turkey, but unlike India it decided to join the US-sponsored defence pacts of the Central Treaty Organization and the South East Asia Treaty Organization when they were formed to create a ring around the Soviet Union in the mid-1950s. The 1965 war with India led to an American embargo on arms sales to Pakistan, which encouraged Pakistan to establish strong links with the People's Republic of China. They subsequently used this link to act as a go-between when President Nixon decided to reverse American policy towards China in the late 1960s.

Throughout, Pakistan's policy was dictated by its fear that India wished to dismember it and then reabsorb the parts into the Indian state. At various stages since 1965 India's other neighbours have shown similar fear or resentment of its power in the region. Nepal has walked the balancing act between its giant neighbours with difficulty, and as the trade disagreements of 1989–90 with India showed, not always with success. After obtaining its independence from Pakistan in 1971 Bangladesh enjoyed a brief honeymoon with India, but that rapidly faded, and for nearly twenty years successive Bangladesh governments have seen themselves as almost powerless pawns in an Indian grip. Until the late 1970s Sri Lanka had enjoyed far more friendly relations with India. The civil war changed that dramatically, for when India saw its own interests threatened by the civil unrest it made it clear to the Sri Lankan government that the pursuit of policies that destabilized India would not be tolerated. Even the tiny island state of the Maldives saw the extent of Indian power in the Indian Ocean when its government called for help to put down an attempted coup in 1988.

# THE GEOGRAPHY OF IGNORANCE

Bhabani Sen Gupta wrote of Afghanistan that 'there is a terrible dearth of reliable credible information, and almost every image is coloured' (Sen Gupta 1986). That statement has far wider validity than to Afghanistan alone, where politically motivated distortion was piled on top of the total lack of objective data gathering. In fact in every region of the South Asian periphery it is essential to emphasize our basic ignorance of many fundamental features of the contemporary geography. Ives and Messerli have shown how widely estimates of 'hard facts', such as those relating to rates of erosion or deforestation in Nepal, vary. Even with all the high technology of satellite imagery and remote sensing it is still impossible to make anything better than general estimates. If that is true for physical data, it is even more so for social and economic information. Certainly for some countries of the region official statistics abound, whether of variables relating to population, the economy, or social characteristics. However, they always have to be treated with great caution. Even the most basic population data are often subject to wide margins of error.

There are enormous problems of analysis and interpretation both through time and from one place to another. Despite the welter of statistical information, there remain great gaps in the information base. Given the nature and complexity of the social and economic geographies of the peripheral states, research on the ground has been remarkably limited. In comparison with many parts of the world, most of South Asia has been described in remarkable detail through a succession of gazetteers and official reports. Yet up-to-date information is geographically patchy and qualitatively very uneven. Analysis often has to be on the basis of evidence taken from very small samples, the reliability and representativeness of which is almost totally unknown. Studies in depth may prove to offer no basis of comparison with areas other than those examined themselves. Thus as Connell and Dasgupta (1978) noted with respect to the myriad village studies in India, the multiple duplication of studies across the sub-continent often appeared to offer little by way of increased depth of understanding.

The problems are exacerbated when change is rapid. The broad outlines of the outer frontiers of geographical knowledge of the South Asian periphery have been described and mapped long back. However, the inner frontiers – the nature and structure of geographical change – often remain virtually uncharted. They remain frontiers of great uncertainty.

## CONCLUSION

Internal diversity has provided the backcloth to the failure of South Asia's peripheral states to develop stable political systems since the mid-1960s. Repeatedly that diversity has both been mirrored in India's own diversity and interacted with it. The choice of contrasting political paths has sometimes sharpened perceived conflicts of interest, but the thread linking the states throughout the period has been the perception of weakness *vis-à-vis* India's regional strength. It is one of the ironies of South Asia that, despite the many common interests that could provide a basis for economic and social co-operation, the last quarter of a century has been marked by repeated conflict both within and between states. It is the more remarkable that despite those conflicts some positive economic change has taken place. Industrial growth and increases in agricultural productivity have helped to keep pace with rapidly growing populations, but the strain on resources, notably those of the environment, appears to be on the increase. In this as in many fields scientific observation is rare and the quality of both information and analysis has not begun to reach the levels demanded by the seriousness of the problems. To carry out that research and to take what steps are necessary to follow it up will take a political willingness to subdue the pursuit of sectarian interest in favour of wider social and political goals of which there is, as yet, little sign.

## REFERENCES

Amjad, R. (1989) *To the Gulf and Back*, Geneva: International Labour Organization.

Anti Slavery Society (1984) *The Chittagong Hill Tracts*, London: Anti Slavery Society.

*Asia Yearbook* (1990) 'Afghanistan' *Far Eastern Economic Review Annual Yearbook*, Hong Kong: *Far Eastern Economic Review*.

Bangladesh Bureau of Statistics (1991) *Statistical Pocket Book of Bangladesh 1990*, Dhaka: Government of Bangladesh.

Bidwai, P. (1987) 'Bhandari Line fits US Plan', *The Times of India* 19 May.

Bradnock, R.W. (1990) *India's Foreign Policy since 1971*, London: Royal Institute of International Affairs/Pinter.

Burki, S.J. (1988) *Pakistan under Bhutto 1971–77*, 2nd edn, London: Macmillan.

Cassen, R.H. (1978) *India: Population, Economy and Society*, London: Macmillan.

Centre for Science and Environment (1985) *The State of India's Environment 1984–85: The Second Citizens Report*, New Delhi: Ravi Chopra Ambassador Press.

Connell, J. and Dasgupta, B. (1978) *Village Surveys in India*, Oxford and Delhi: University of Sussex.

Farmer, B.H. (1983) *An Introduction to South Asia*, London: Methuen.

Ives, J. and Messerli, B. (1989) *The Himalayan Dilemma*, London and New York: The United Nations University and Routledge.

Jayawardene, K. (1987) *Ethnic and Class Conflict in Sri Lanka*, Madras: Kaanthalakan.

Johnson, B.L.C (1979) *Pakistan*, London: Heinemann.

Johnson, B.L.C (1982) *Bangladesh*, London: Heinemann

Johnson, B.L.C and Scrivenor, M. Le M. (1981) *Sri Lanka: Land, People and Economy*, London: Heinemann.

Khoshoo, T.N. (1986) *Environmental Priorities in India and Sustainable Development*, New Delhi: Presidential Address, India Science Congress Association.

Muni, S.D. and Muni, A. (1984) *Regional Co-operation in South Asia*, New Delhi: National Publishing House.

Government of Pakistan (1988) *Pakistan Yearbook 1986–87*, Islamabad: Government of Pakistan.

Rose, L.E. (1989) 'India and the world', in Robinson, F. (ed.) *Cambridge Encyclopedia of India, Pakistan, Bangladesh and Sri Lanka*, Cambridge: Cambridge University Press.

Schwartz, W. (1987) *Sri Lanka's Tamils*, London: Minority Rights Group.

Sen Gupta, B. (1986) *Afghanistan*, London: Pinter.

de Silva, K.M. (1981) *A History of Sri Lanka*, Oxford: Oxford University Press.

Singer, M.R. (1991) 'Sri Lanka in 1990: the ethnic strife continues', *Asian Survey* 31(2): 140–5.

Swan, B. (1987) *Sri Lankan Mosaic: Environment, Man, Continuity and Change*, Colombo: Marga Institute.

World Bank (1989) *Bangladesh Action Plan for Flood Control*, Washington DC: World Bank.

World Bank (1991) *The World Development Report*, London: Oxford University Press.

Ziring, L. (1991) 'Pakistan in 1990: the fall of Benazir Bhutto', *Asian Survey* 31(2): 113–24.

# 4

# THE RISE OF THE NAGA
## The changing geography of South-East Asia 1965–90

*Jonathan Rigg and Philip Stott*

## INTRODUCTION: SOUTH-EAST ASIA IN 1965

In 1965 the Department of Geography at the School of Oriental and African Studies, in the University of London, was founded by Professor C.A. Fisher, who had recently completed a monumental survey of the social, economic and political geography of South-East Asia, a study which was to become the benchmark for future work in the region (Fisher 1964). At the time he was writing, the prognosis for this distinctive tropical, maritime realm of newly independent states was far from certain:

> it must be hoped that the newly independent states will succeed in stamping out the widespread lawlessness and corruption which still remain in some of them as legacies of the momentous upheavals of the 1940s. For only if this is done will it be possible effectively to raise the living standards of their peoples, to check the drift to further Balkanization, and to prevent the region from becoming, as at times has seemed only too probable, the powder keg of Asia.
>
> (Fisher 1966: 10)

Early freedom movements, later encouraged by Japanese occupation during the Second World War, had helped to create a climate for independence from American, British, Dutch and French administrations during the post-war period, so that by 1965 most of the modern states of South-East Asia had emerged from the colonial cocoon: the Philippines in 1946, Burma (now Myanmar) in 1948, Indonesia in 1949, Laos and Cambodia in 1954, and Malaysia in the period from 1957 to 1963 (see Figure 4.1). In 1965, when

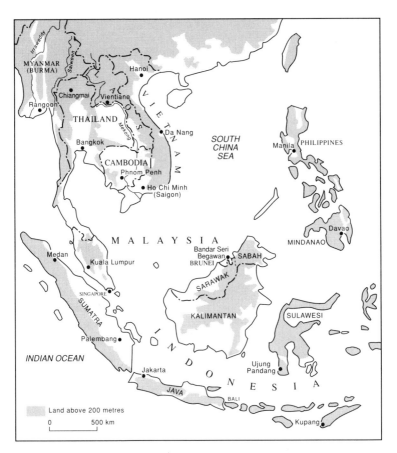

*Figure 4.1* The states of South-East Asia

Lee Kuan Yew proved too difficult a partner for the wider federation of Malay states, Singapore was forced out of Malaysia to become yet another independent state. In contrast, the small, but immensely oil-rich, Sultanate of Brunei, situated on the north-west coast of Borneo, never became part of Malaysia, but it did cease to be a British Protectorate in 1971, although full independence was not achieved until 1983, when British troops were finally withdrawn, and the Sultanate became responsible for its own defence.

The final independence and unity of Vietnam, however, was to take a longer and bloodier road than that followed elsewhere in

75

South-East Asia, being eventually achieved in 1975–6 under a communist government, after some thirty years of war and international involvement. In 1954, following a shattering defeat of the French at Dien Bien Phu, the country had been divided along the seventeenth parallel. By 1960, the North Vietnamese, supported by the People's Republic of China, were sending arms and troops in support of the Viet Minh, the communist guerrillas in the South of the divided country, and by 1965, there was full-scale war, with the forces of the United States of America committed to saving the South Vietnamese regime. At last, in April 1975, South Vietnam fell to the North, and the country was finally reunified, in 1976, as the Socialist Republic of Vietnam, with Saigon renamed Ho Chi Minh City.

The year 1965 was thus a key turning point in the development of modern South-East Asia: the colonial days were over, Singapore was just created, and the most dangerous period of the Vietnam conflict was about to start, a period that would see both newly independent Cambodia and Laos dragged into the theatre of war. At that time, many commentators earnestly believed that other South-East Asian countries, especially Thailand – the one nation in South-East Asia never to have been ruled directly by a colonial power – would soon be engulfed by the conflict. In stark contrast, following its declaration of the 'Burmese Way to Socialism' in 1962, Burma (now Myanmar) was already idiosyncratically isolated, and its economy on the path to stagnation. The future for the peoples of the region, who still inhabited a land primarily characterized by traditional agriculture and tropical forests, did not look good, and there seemed little chance for 'a reasonable space of time', in which South-East Asia could 'work out its own answers to the many and pressing internal problems' with which it was faced (Fisher 1966: 777). Yet, in 1967, a first step was made with the establishment of the Association of South-East Asian Nations (ASEAN), a move which would soon divide the region into two different realms.

## SOUTH-EAST ASIA IN 1990

The South-East Asia of the 1990s is radically different from that of 1965. To begin with, political and economic cleavages which stem essentially from that date, have led, arguably, to the emergence of three 'South-East Asias': the market economies of the Association of South-East Asian Nations (ASEAN), comprising Brunei, Indonesia,

Malaysia, the Philippines, Singapore and Thailand; the command economies of Indochina, namely Cambodia, Laos and Vietnam; and idiosyncratic Myanmar (formerly Burma) (see Figure 4.1).

For the countries of ASEAN, developments over the period since 1965 can be viewed with relative satisfaction. Weighted annual economic growth rates have averaged 6.3 per cent. There is talk, admittedly often unfocused and ill-defined, of Malaysia and Thailand joining Singapore in the exclusive club of 'newly industrializing countries' (NICs). Politically there has also been a gradual movement towards more representative, democratic government, most visibly demonstrated when Ferdinand Marcos was forced to relinquish the presidency of the Philippines in 1986.

The countries of Indo-China have less to be sanguine about. Mirroring developments in Eastern Europe and the former Soviet Union, there has been a loss of confidence in the socialist road to reconstruction and development. Relative to the market economies of the region, and even given the dearth of accurate data available, it is clear that incomes have remained stagnant, pressurizing the leaderships in each to implement economic reforms: private markets, market-based incentives, and greater foreign investment (Table 4.1). They are admitting to serious levels of underemployment and unemployment, and countenancing the possibility of bankruptcies. Even the government of Myanmar has been forced by economic necessity partially to reject the economic autarky of the previous two and a half decades, and to establish joint ventures with foreign companies. Politically, signs of 'progress' in Myanmar and the countries of Indo-China are rather more limited. Worried about events in the wider world, the leaderships are attempting the difficult task of promoting economic freedom, but at the same time stifling political freedom. The examples of Tiananmen Square and Eastern Europe illustrate the potential risks of such a strategy, both to the governments and to their peoples.

Among the countries of ASEAN, the economic changes of the past twenty-five years have taken the pattern of those in many other developing countries, only at an even faster pace. Industry, and particularly manufacturing, has grown at an impressive rate, while agriculture has done so more slowly (Table 4.1). In some cases, farmers have been encouraged to leave the countryside in search of better paid employment in the cities, contributing at the same time to a burgeoning urban population and an increase in the amount of idle agricultural land. The promotion of rapid growth

Table 4.1 Population, land area and economic growth in South-East Asia

| | Land area (sq km) | Income per capita 1988 (US$) | Average annual growth rate of GDP (%) | | Average annual growth rate 1980–8 (%) | | Employment in Agriculture (% total) | Average annual growth rate in exports (%) | |
|---|---|---|---|---|---|---|---|---|---|
| | | | 1965–80 | 1980–8 | Agriculture | Industry 1987–88 | | 1965–80 | 1980–8 |
| **ASEAN** | | | | | | | | | |
| Thailand | 513,100 | 1,040 | 7.2 | 6.0 | 3.7 | 6.6 | 58 | 8.5 | 11.3 |
| Malaysia | 330,400 | 1,880 | 7.3 | 4.6 | 3.7 | 6.1 | 31 | 4.4 | 9.4 |
| Singapore | 620 | 8,160 | 10.1 | 5.7 | -5.1 | 4.5 | 1 | 4.7 | 7.3 |
| Brunei | 5,800 | 12,770[a] | n.a. | n.a. | n.a. | n.a. | 22[b] | n.a. | n.a. |
| Indonesia | 1,919,000 | 400 | 8.0 | 5.1 | 3.1 | 5.1 | 55 | 9.6 | 2.9 |
| Philippines | 300,000 | 670 | 5.9 | 0.1 | 1.8 | -1.8 | 48 | 4.7 | 0.4 |
| **Indo–China and Myanmar** | | | | | | | | | |
| Vietnam | 329,600 | 100–150 | n.a. | n.a. | n.a. | n.a. | 81[c] | n.a. | n.a. |
| Laos | 236,800 | 160[d] | n.a. | n.a. | n.a. | n.a. | 90[e] | n.a. | n.a. |
| Cambodia | 181,000 | n.a. | n.a. | n.a. | n.a. | n.a. | n.a. | n.a. | n.a. |
| Myanmar | 676,500 | 210 | n.a. | n.a. | n.a. | n.a. | 65 | -2.1 | -7.0 |

Sources: *Asia Yearbook* (1990); World Bank (1990).
Notes: [a] 1987
   [b] 1986
   [c] 1979: population living in the countryside.
   [d] 1987
   [e] 1985: population living in the countryside.

and 'Western-style' development, sometimes at any cost, has also brought with it serious environmental problems. The forests of South-East Asia have been decimated to gain valuable foreign exchange, while levels of pollution in the cities and sometimes in the countryside are worryingly high. The recognition that economic development should not be at the expense of the environment has spawned a powerful environmental lobby.

Certain developments in the geography of South-East Asia seem fairly clear: the market-based strategy of economic development among the countries of ASEAN for example. Other episodes and developments are, seemingly, only nascent. How will the countries of Indo-China respond to the economic necessity of improving growth rates and living conditions? How will this impinge upon the political system? Will there be a political fragmentation on the scale of that in Eastern Europe (not an occurrence that could be easily predicted)? All that can be attempted in an account such as this is to sketch out the past and the future in the manner which makes most sense at the time of writing.

## POPULATION GROWTH

In 1965 the countries of South-East Asia had a combined population of 247 million. By 1989 this had increased to 444 million (Table 4.2). Although the region remains relatively land rich when compared with the neighbouring countries of East and South Asia, many officials now accept that they are facing a population 'problem', requiring the introduction of policies of population control. Officials and academics usually express the view that the countries in question do not have the physical and economic resources to absorb a rapidly growing population – and that this, in essence, is the problem with which they have to deal. Limited land means that the numbers of land poor and landless farmers will increase, industry will be unable to absorb the 'surplus' population, and the government will not be in a position to provide adequate levels of, for example, health care and education. In short, excessive population growth will retard economic and social development.

However the issue is not simply one of *overpopulation*, or at least imminent overpopulation. Many would view it more accurately as one of *maldistribution* and *underproduction*. Thus, often in conjunction with family planning programmes, governments in the region have also implemented extensive schemes to redistribute

*Table 4.2* South-East Asian demography

| | Population | | Crude birth rate per 1,000 population | | Crude death rate per 1,000 population | | Total fertility rate | |
|---|---|---|---|---|---|---|---|---|
| | 1965 | 1989 | 1965 | 1988 | 1965 | 1988 | 1965 | 1988 |
| ASEAN | | | | | | | | |
| Thailand | 31.4 | 55.6 | 41 | 22 | 10 | 7 | 6.3 | 2.5 |
| Malaysia | 9.5 | 17.4 | 40 | 30 | 12 | 5 | 6.3 | 3.7 |
| Singapore | 1.9 | 2.7 | 31 | 18 | 6 | 5 | 4.7 | 1.9 |
| Brunei | 0.3 | | | | | | | |
| Indonesia | 184.6 | 184.6 | 43 | 28 | 20 | 9 | 5.5 | 3.4 |
| Philippines | 31.8 | 64.9 | 42 | 31 | 12 | 7 | 6.8 | 3.8 |
| Indo-China and Myanmar | | | | | | | | |
| Vietnam | 35.1 | 66.8 | n.a. | 31 | n.a. | 7 | n.a. | 4.0 |
| Laos | 2.7 | 3.9 | 45 | 47 | 23 | 17 | 6.1 | 6.6 |
| Cambodia | 6.1 | 6.8 | 44 | n.a. | 20 | n.a. | 6.2 | n.a. |
| Myanmar | 24.3 | 40.8 | 40 | 30 | 18 | 10 | 5.8 | 3.9 |

*Source:* World Bank (1990).

people from over- to underpopulated areas, as well as encouraging increased production in those areas deemed to be overpopulated. To complicate the issue still further, the issue of population growth and overpopulation is also a political issue – not just one of demographics. The settling of frontier areas in Indonesia with Javanese transmigrants (see pp. 83–4) is seen by some as an attempt to secure what are perceived to be 'sensitive' areas populated by dissatisfied ethnic minorities. While in Malaysia, the government's pro-natalist policy pursued since the early 1980s has been driven, in part, by a concern that the bare Malay, or *bumiputra*, majority (58 per cent of the total in Peninsular Malaysia) be maintained. If current levels of fertility among the Chinese and Malay population persist in Malaysia, then the *bumiputra* share will increase to 62 per cent in the year 2000, and to 70 per cent by 2025 (Bauer and Mason 1990: 44). A target population of 70 million has now been set.

## Family planning in Thailand

As a whole, South-East Asia is at an intermediate stage in the demographic transition. Mortality rates have declined significantly, and fertility rates are in the process of following (Table 4.2). Singapore, Malaysia, Thailand, Vietnam, perhaps Burma, and some areas of the Philippines and Indonesia have witnessed impressive falls in fertility – a trend which appears as though it will continue. However, governments have, in the main, not been willing to let this occur 'naturally'. Programmes of family planning – population control – have also been initiated to accelerate the process. Possibly the most innovative and successful such programme has been that introduced by the Thai authorities. The success is especially noteworthy in that it has been achieved over a relatively short period of time and amongst a largely rural, agricultural population.

The Thai government did not adopt an official population policy until March 1970. From the end of the war until 1961, Thailand's policy was essentially pro-natalist, and population growth rates exceeded 3 per cent per year. Following a World Bank mission to Thailand in 1957–8, which linked the high rates of population increase with a possible slow-down in economic growth, the government did begin – albeit hesitantly – to give advice on birth control. This eventually led to the implementation of an overt policy of population control in 1970. The Thai Cabinet explained their actions as follows:

> The Thai government has the policy of supporting family planning through a voluntary system, in order to resolve serious problems concerned with the very high rate of population increase which constitutes an important obstacle to economic and social development of the nation.
>
> (quoted in Penporn Tirasawat 1984: 112)

The Ministry of Public Health was instructed to establish a National Family Planning Project, and in 1974 these efforts were intensified in rural areas when the Community Based Family Planning Service (CBFPS), a private non-profit development organization, was established and placed under the leadership of the charismatic Mechai Viravaidya (Krannich and Krannich 1980). Successive National Economic and Social Development Plans have aimed for a reduction in the population growth rate from 3.0 per cent in 1972, to 2.5 per cent in 1976, to 2.1 per cent in 1981, and to 1.5 per cent by 1986. The Sixth National Economic and Social Development Plan (1987–91) set out to reduce population growth still further, to 1.3 per cent by 1991 (NESDB n.d.: 77). Remarkably for any government plan, actual population growth rates have closely mirrored those targeted in the plans, and the Thai experience is usually perceived to have been a great success (see Figure 4.2).

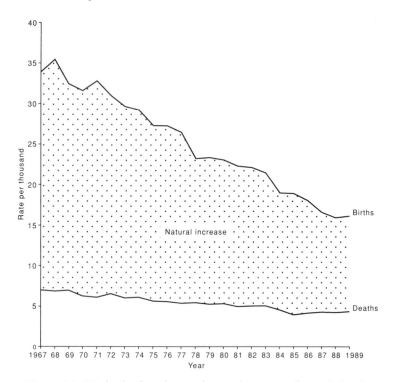

*Figure 4.2* Birth, death and rate of natural increase of population in Thailand 1969–87

This success can be put down to two main factors. First, Mechai Viravaidya recognized that he should focus upon the cultural constraints to adoption. To this end, a 'road show', with free coloured condoms, T-shirts and condom-blowing contests helped to desensitize an issue that most Thais were not used to discussing, and certainly not with strangers (condoms even came to be known as 'mechais'). At the same time, the programme managed to garner the support of the *sangha* or monkhood, using Buddhist texts to support the notion of planned parenthood (e.g. 'many births cause suffering') and getting monks to bless contraceptive pills and condoms.

The second principal element in the successful recipe was the concentration of resources within the programme. As a result there was no need to elicit the support of other departments or ministries, and although this led to a degree of duplication of resources, it did enable the programme to be implemented with impressive speed and, in the case of the CBFPS, to penetrate remote rural areas where the government's own health service was weak. The community-based approach to family planning, and its integration with wider issues of development (the CBFPS provided cheap bank loans for example), also encouraged a positive response from the farmers. The spread of both knowledge about family planning and methods of family planning was rapid and pervasive, to the extent that many commentators talked of a 'reproductive revolution' (Rosenfield et al. 1982: 43).[1]

## Land settlement in Indonesia: the transmigration scheme

Indonesia, like Thailand, has also developed a comprehensive family planning programme. However, in Indonesia's case, far greater resources have been directed towards the redistribution of popula-tion. Of Indonesia's total population of 185 million, 107.5 million, or almost 60 per cent, are concentrated on the four metropolitan islands of Java, Madura, Bali and Lombok – which account for only 7 per cent of the total land area. Agricultural population densities exceed over 1,000 per sq km over large areas of Java and Bali, while in the 'Outer Islands' (Sumatra, Kalimantan, Sulawesi and Irian Jaya) they are often less than 1 per sq km.[2] The uneven distribution of population led the Indonesian government to build upon previous Dutch initiatives and initiate the transmigration programme in 1950. Sukarno deemed it 'a matter of life and death for the Indonesian nation' (Caufield 1985: 188). Since then, approaching 6.5 million people have been resettled in the Outer Islands, and the programme has consumed, at times, over 6 per cent of the total development budget (Arndt 1984: 40; Hardjono 1986; World Bank 1988). Although the programme has had multiple objectives that have changed in their relative importance over time (regional development, security, food production and improve-ments in general welfare) the demographic rationale – the solving of Indonesia's population 'imbalance' – has remained deeply rooted.

From its inception, the scheme has attracted widespread criticism on two principal fronts: the practical or operational, and the fundamental. The former has tended to stress the selection of inappropriate migrants in terms of age, family size or skills; the selection of inappropriate or poorly prepared sites; the failure to develop farming systems appropriate to environmental conditions in the Outer Islands; and the related issues of poor management and administration. As the programme has been modified and improved, so some – although by no means all – of these difficulties have been overcome (see e.g. Hardjono 1986; 1988; Arndt 1983). The second area of criticism however is far more intractable, for it focuses on the *raison d'être* of transmigration and its impact on settlement areas. Critics maintain that the scheme has displaced tribal peoples in Kalimantan and Irian Jaya, resulting in a 'Javanization' of the Outer Islands and the consequent erosion of local cultures leading, in some cases to 'ethnocide' (Colchester 1986). Further, it has led to the destruction of one of the most valuable natural resources in the world – the tropical rain forest ecosystem (see pp. 96–102). For those individuals who stress such 'fundamental' problems (e.g. Otten 1986; Colchester 1986), the transmigration scheme represents an insidious programme of social engineering which cannot be improved, and should be curtailed.

However, the most important lesson to be learnt from the experience of the transmigration programme, and this also applies to resettlement schemes in Vietnam, Cambodia, Malaysia, Thailand and the Philippines, is that population 'imbalances' in South-East Asia are rarely the result of historical accident: they usually reflect variations in agricultural potential. Today, the reality is that there are few large tracts of good agricultural land left in the region, and little chance that surplus population can be accommodated through its redistribution. The solution to impending overpopulation must lie in a reduction in population growth rates coupled with increased production, both agricultural and industrial.

## PLURAL SOCIETIES

If maldistribution and underproduction are seen as key population issues in South-East Asia, the *cultural diversity* of the region is just as significant, for all the modern states of South-East Asia are plural societies. In fact, the first use of the term 'plural' in this context was by Furnivall (1948; 1980), who was writing of Burma (now Myanmar) and the Netherlands Indies (Indonesia), and thus the whole concept is South-East Asian in character. Today, the idea has become especially associated with the discussion of the relationships between ethnic Chinese and the indigenous population in countries such as Malaysia, largely because the Chinese tend to comprise the biggest and often the most successful minority group (Table 4.3). In South-East Asia as a whole, they represent around 7 per cent of the total population, with a range from 76 per cent in the case of Singapore to less than 2 per cent in the Philippines. Needless to say, all such figures are very much rough estimates, because the Chinese population is often difficult to identify accurately. In Thailand, for example, many 'Thais' are Chinese to various degrees, and here estimates have varied from 8.7 per cent to 15 per cent.

*Table 4.3* The Chinese in South-East Asia

| Country | Date | Number | Percentage of total population |
|---------|------|--------|-------------------------------|
| Brunei | 1980 | 54,150 | 25.4 |
| Indonesia | 1980 | 4,116,000 | 2.8 |
| Malaysia | 1985 | 4,882,300 | 30.9 |
| Philippines | 1981 | 699,000 | 1.5 |
| Singapore | 1989 | 2,038,000 | 75.9 |
| Thailand | 1980 | 6,000,000 | 13.0 |
| ASEAN | 1981 | 16,939,171 | 6.6 |

*Source:* Adapted from Rigg (1991a: Table 6.1 and p. 110).

However, the essential plurality is much wider than simply the Chinese presence. In religious terms, South-East Asia remains divided between Islam (Brunei, Malaysia and Indonesia), Buddhism (the mainland states) and Roman Catholicism (the Philippines), with animism, Hinduism (especially in Bali) and Brahmanism underlying the beliefs and practices of the whole region. As on the border between Malaysia and Thailand (Islam/Buddhism) and in southern Mindanao in the Philippines (Islam/Roman Catholicism), zones of friction still exist where the great religions come into geographical proximity, and such tension points could become increasingly significant following any rise in Islamic fundamentalism, although Islam in South-East Asia is still primarily influenced by local factors, and most other Islamic societies regard it as somewhat unorthodox. Every country in the region is also the home of a wide range of indigenous peoples, who have arrived at different times, and who have waxed and waned in their relative dominance. In Indonesia, for example, some 25 major languages are spoken, with over 250 dialects. The most numerous peoples here are the Javanese, Sundanese, Madurese, Malays, Bataks, Buginese and Balinese; there remain also Indian, Arab, Chinese and European communities, many well-established after several generations of settlement. One measure of unity adopted by the Indonesian government has been to make *Bahasa* ('market') Indonesian the *lingua franca* of the country, and the government has further agreed with its counterpart in Malaysia on a common Roman script, thus making this the official language of both states.

Ethnic tensions are often particularly seen in the context of the access to and exploitation of land and other resources. In Thailand, for example, the shifting cultivators (swiddeners) of the uplands, such as the Hmong and Karen, are often competing for important watersheds with Thai lowlanders (*khon muang*), who are themselves expanding into the mountains, or who need to employ the water supplies from the hills for their own lowland irrigation systems. Competition and friction also exist over the use of forest resources and wildlife, and through the role of hill peoples in the production of opium in the 'golden triangle' region of Myanmar, Laos and Thailand. It is a common theme that there has been a long history of misunderstanding and distrust between the dominant lowlanders of South-East Asia, such as the Burmans, the Thais and the Vietnamese, and the peoples who inhabit the mountains, groups who often fail to recognize the modern boundaries of the independent nation states. In Myanmar especially, insurgency continues, with several communist, non-communist and ethnic rebel groups, such as the Shans (around 8.5 per cent of the population), active. Throughout the region, shifting cultivators are blamed for the destruction of watersheds, soil erosion and the loss of wildlife, even when the real causes clearly lie elsewhere, often indeed with the very governments concerned.

In Malaysia, the problems of the plural society have had a particular legal and political expression in the form of the New Economic Policy (NEP), the establishment of which in 1971 followed the July 1969 race riots, in which 235 people were officially recorded as dead or missing, although the real figure is probably much higher (Strauch 1981). One of the main aims of the plan, which was a growth-orientated strategy for the elimination of poverty and not just a redistributive operation, was, nevertheless, to restructure Malaysian society in order 'to correct economic imbalance, so as to reduce and eventually eliminate the identification of race with economic function' (Government of Malaysia 1971). This policy was to be achieved over a twenty-year period, that is by 1990, through the development of a labour force which mirrored the overall ethnic composition of the country, and through the restructuring of business ownership, so that Malays and other indigenous, non-Chinese, groups (the so-called *bumiputras*, 'princes' or 'sons of the soil', mentioned earlier on p. 80) managed at least 30 per cent of the share capital of the corporate sector. We have now reached 1990, and these principles are still very much a part of Malaysian life, so that Rigg (1991a: 121) is able to write, 'the NEP is a policy that remains central to the operation of the Malaysian political economy.' In fact, the *bumiputra* policy has somewhat institutionalized the significance of ethnic divisions, and Malays continue to distinguish themselves from the Chinese as economically and educationally deprived. 'Malaysians' have therefore not yet supplanted the more ancient ethnic pluralities of Malays, Chinese, and others.

The long conflict over Indo-China also continues to hinder the development of South-East Asia, with Cambodian refugee camps dotted along the Thai side of the Cambodian border, causing a new plurality and severe problems for Thailand. For three years, from 1976 until 1979, Cambodia was ruled, as Democratic Kampuchea, by the Khmer Rouge under Pol Pot, alias Saloth Sar; during this period, the population of the country was reduced by some 2 million through brutal murder, disease and starvation. In January 1979 Pol Pot's horrific regime was replaced by the Vietnamese-backed People's Republic of Kampuchea. The Vietnamese withdrew in 1989, but it was only in 1991 that a comprehensive peace settlement was finally agreed – whether it will be successfully implemented, with UN supervision, is yet to be seen.

Ethnic, religious and cultural plurality, both within and between states, thus continues as a hallmark of modern South-East Asian societies, and it is not always easy to know precisely when local skirmishes may develop into something more significant. One essential criterion defining South-East Asia as a region, therefore, remains its own intrinsic cultural diversity, a theme stressed by Fisher in his *Geography* of 1964.

## ECONOMIC CHANGE

### Agricultural change

Despite rapid and far-reaching structural changes in the economies of the countries of South-East Asia, the bulk of the population of the region still live in the countryside, and earn their livelihoods from agriculture (Table 4.1). It might be assumed that agriculture has changed little, for in many areas the countryside appears – at least superficially – to conform to the traditional ideal of rural life. However, such a perspective overlooks a number of less obvious but none the less pervasive changes which have fundamentally altered the nature of farming and rural life in the region.

Clearly, the market and command economies of South-East Asia have followed radically different paths to development, and it is therefore difficult to examine the two groups of countries together, as a single unit. Because of the general dearth of data – and particularly of village-level studies – from the countries of Indo-China and Myanmar, much of the following discussion will focus on conditions in ASEAN. None the less, there are parallels between the two groups of countries. Among the market economies, many of the changes can be linked to the processes of commercialization and commodification. And among the command economies, to the intrusion of the state in the guise, particularly, of communist and socialist parties.

The commercialization of production and life in South-East Asia has altered farmers' lifestyles, their aspirations and, as a result, their strategies of production. Traditionally, and excepting such areas of export production as Lower Myanmar, the Chao Phraya Delta of Thailand, Java, and the Mekong Delta, farmers secured the resources for cultivation locally, and production was geared towards meeting the subsistence needs of the individual household. Surpluses were limited, markets were often inaccessible, and strategies of production were geared towards the minimization of risk, not the maximization of profit. However, with improving communications, a widening 'pressure of needs', and the intrusion of the cash economy into most areas of South-East Asia, so farming has become correspondingly commercialized. Farmers now *need* to produce a surplus to be sold for cash to meet the costs of the goods (radios, TVs, cold drinks) and services (education, health care, electricity) which are an essential part of any respectable family's lifestyle. To this end, farmers are growing different crops and crop varieties using new technology. At the same time, the processes of agrarian change have also altered the ownership, distribution and use of land and of labour.

## Crops and technology

The most far-reaching technological change has been associated with the Green Revolution in rice production – the pre-eminent staple crop of the region (for a general review see Barker et al. 1985; Rigg 1989). The Green Revolution embodies the use of new high yielding or modern varieties of rice (MVs) and their cultivation with large quantities of chemical fertilizers, pesticides and herbicides. Some commentators would also include mechanization and irrigation as elements of the Green Revolution 'package'.

All the countries of South-East Asia, the command as well as the market economies, have embraced this new rice technology (NRT) (Table 4.4). Indeed, arguably the origins of the Green Revolution lie in Vietnam, where early-maturing (60–100 days after transplanting) Champa rices were discovered and then disseminated into southern China around 1000 AD (Dalrymple 1986:15). However, the country which has adopted the NRT with the greatest enthusiasm is the Philippines where, notably, the International Rice Research Institute (IRRI) is located (see Barker 1985). A study by Herdt (1987) which reviewed the changes in the technology of rice farming in central Luzon and Laguna between 1965 and 1982 illustrates the degree of change. In 1965–6 no farmers used MVs, rates of fertilizer application stood at 20 kg/ha, while expenditure on herbicides and pesticides averaged 3–5 pesos/ha. In 1981–2, virtually 100 per cent of farmers used MVs, fertilizer applications averaged 78–83 kg/ha, and expenditure on herbicides and pesticides 151–6 pesos/ha. Over the same time, the use of power tillers on fields increased from 0 per

Table 4.4 Proportion of riceland planted to modern varieties of rice in South-East Asia

| ASEAN | |
|---|---|
| Indonesia | 54% (1984) |
| Brunei | n.a. (but cultivated) |
| Malaysia | 54% (1982) |
| Philippines | 85% (1983) |
| Thailand | 13% (1982) |
| **Indo-China and Myanmar** | |
| Myanmar | 49% (1984) |
| Cambodia | n.a. (but cultivated) |
| Laos | n.a. (but cultivated) |
| Vietnam | 32% (estimate, early 1980s) |

Source: Dalrymple (1986).

cent to 58 per cent in the case of central Luzon, and from 38 per cent to 98 per cent in the case of Laguna.

Perhaps the most contentious issue is the distributional effects of the new technology. Who has benefited from the Green Revolution? Although there is not space here to review comprehensibly the debate, there has been a discernible shift away from the view that holds that only wealthier, larger, landowners have benefited, to one that maintains that rich and poor alike have found attractions in the new technology (see Barker et al. 1985; Rigg 1989; 1990; Lipton and Longhurst 1989; for alternative viewpoint see Cederroth and Gerdin 1986; Pearse 1980; Feder 1983; Hüsken 1979). Further, the technology is becoming increasingly 'small-holder friendly' as it develops, and in Indonesia, Malaysia and the Philippines over 80 per cent of farmers are now planting MVs. To proponents of the new technology in South-East Asia they see 'a real danger of growing inequality in rural areas not because of the MV technology, but because of insufficient progress in the technology under the strong population pressure on land' (Hayami 1984: 394).

## Labour and land

The processes of agrarian change, driven by the commercialization of production and life, have also fundamentally altered the management of land and labour. Land is now an economic resource that can be bought, rented and sold. Communal land is becoming less common, tenancy and land fragmentation are increasing, and the pattern of land ownership is becoming skewed towards a rich minority of large land owners. In 1988 in the province of Ayutthaya in the Central Plains region of Thailand, the amount of land rented amounted to 53 per cent of total farmland (MOAC 1989: 220–1). In Hüsken's study of the village of Gondasari in Java, he records that two-thirds of households depend for their livelihoods on working the land of others, that the position of the landless has progressively deteriorated, and that this has been in large part due to the Green Revolution and associated commercialization. He notes how the landless are now at the mercy of the land owners and quotes a farmer as saying 'the little men are all beaten to death' and 'the big water buffaloes can win because they use their horns' (Hüsken 1979: 146). The rise in landlessness in the countries of South-East Asia has created a large dispossessed and dissatisfied peasantry. In Thailand during the late 1970s and early 1980s when the

Communist Party of Thailand was influential, and in the Philippines today where the New People's Army fields an estimated 20,000 rebels (and where 30–5 per cent of the agricultural population are landless), the leaders of the insurgencies have successfully played on the emotive issue of land reform.

However, this rise in tenancy, landlessness and land fragmentation is not purely due to the pressures of commercialization. Population growth in areas where land is a resource in short supply (in certain areas the stock of agricultural land is actually decreasing as urban areas expand) has led to a dramatic fall in available land. In Java as a whole, the per capita area of *sawah* (riceland) has decreased from 0.070 ha in 1940, to 0.059 ha in 1955, to 0.038 ha by 1980 (Booth 1985: 123).

Particularly striking is the extent to which traditional forms of reciprocal labour exchange (*long khaek* in Thailand, *gotong rojong* in Java, *berderau* in Malaysia) are being replaced by wage labouring, adapted forms of exchange, or by the mechanization of production. In Indonesia, there has been great attention paid to the substitution of the traditional *bawon* system of rice harvesting in which all villagers had a right to engage (sometimes there might be 500 harvesting a single field), by the *tebasan* system in which the standing crop is sold to a middleman, thus bypassing the villagers and eliminating their right to be involved in the harvest. Although the implications of the change, and even the change itself, are still disputed, it is commonly felt that it represents just another instance in which the commercialization of production has led to benefits accruing to a wealthy minority at the expense of the community as a whole (see Collier et al. 1973; 1974; Collier 1981; Alexander and Alexander 1982; Schweizer 1987).

Mechanization has also apparently served to undermine the position of the poorer segments of village society who depend for their livelihoods on working the fields of others. Mechanization is seen to have displaced labour, a process which is perhaps most vividly illustrated in Scott's study of a village in the Muda irrigation scheme of Peninsular Malaysia. Of the spread of combine harvesters, he writes:

It is hard to imagine the visual impact on the peasantry of this mind boggling technological leap from sickles and threshing tubs to clanking behemoths with thirty-two-foot cutting bars. ... the combines have cut paddy wage labour by 44 per

91

cent. ... Mechanisation, by promoting large-scale farming and leaseholding, has greatly reduced the opportunity for small-scale tenancy. It has also eliminated gleaning, shifted local hiring patterns, reduced transplanting wages, and transformed local social relations.

(Scott 1985: 75–6)

## Industrial change

Agricultural changes in South-East Asia have at least been wrought within the context of a traditional occupation. Industrial growth and development has, in most instances, led to the transplantation of entirely new activities. This has obviously been underway from the period of colonial rule, but in the last quarter of a century the process has accelerated and intensified. Two developments stand out particularly starkly. First, the switch among the countries of ASEAN towards export-orientated strategies of industrialization (EOI), and second, the economic reforms that the countries of Indo-China have been forced to implement.

## EXPORT-ORIENTATED INDUSTRIALIZATION AMONG THE ASEAN NATIONS

In 1965 all the countries of South-East Asia were following strategies of import substitution industrialization (ISI). Imports were restricted through the use of tariffs and quotas, thus protecting domestic 'infant' industries and encouraging domestic companies to substitute for previously imported goods. These policies led to healthy rates of growth during the early years of independence. However, by the 1970s (earlier in Singapore) it was becoming clear to some local economists and ministers that the scope for ISI was gradually diminishing as industrialization proceeded. Domestic markets had become saturated, and the protection afforded to domestic manufacturers had led to the growth of small and often highly inefficient enterprises. Further, the close involvement of governments in the region in industrial matters – the granting of licences to prospective manufacturers for example – had led to a surge in what has been termed 'rent seeking': in short, corruption. In the Philippines this became known as 'crony capitalism', describing the manner in which families and individuals gained concessions and other business advantages from their close association with President Marcos. Crony capitalism is endemic among the market economies of South-East Asia (with the notable exception of Singapore), and indeed, it is virtually impossible to succeed in business without cultivating such links (see Clad 1989; Yoshihara 1988).

However, of all the factors which encouraged the governments of ASEAN to turn away from ISI towards a more export-orientated strategy the most compelling was the success of the so-called newly industrializing countries (NICs). By promoting exports, South Korea, Taiwan, Hong Kong and Singapore (the last obviously a South-East Asian country) attained spectacularly high rates of economic growth. Between 1965 and 1980 Hong Kong and South Korea's GDP, for example, expanded at an average annual rate of 8.6 per cent and 9.6 per cent respectively, and between 1980 and 1988 at 7.3 per cent and 9.9 per cent (World Bank 1990). With the logic of ISI waning, the market economies of South-East Asia followed suit, each turning to EOI during the 1970s and 1980s. If simple growth in manufactured exports is any indication, then the shift to EOI has been successful (Table 4.1). However, it is important to stress that it is only a shift, not a wholesale change. Each of the countries continues to protect inefficient domestic manufacturers, and crony capitalism and rent-seeking are still widespread. But there has been a change in emphasis towards reducing tariffs rates and quotas, and promoting internationally competitive industries. There are risks involved in such a strategy, for it makes the countries in question far more dependent on developments beyond their control – in the international market place (see Rodan 1989 on Singapore).

This has led numerous commentators in recent years to ask whether these countries – and in particular Thailand and Malaysia – represent a second tier of NICs (e.g. *Newsweek* 1988; *The Economist* 1989; FEER 1989). Such a view is tempting, but premature. The first point is that it is far from clear what constitutes a NIC. More importantly, although industrial growth may be impressive, Malaysia, Thailand, Indonesia and the Philippines still all have large and poor rural populations. Disparities in incomes between rural and urban areas have widened during the course of the 1970s and 1980s and until they narrow it is difficult to argue that NIC-dom has been attained. In

Thailand, average incomes in the poor, rural, North-east were 2,961 baht/household in 1988 and the wages of a farm labourer are likely to be in the region of 30 baht (£0.60). In Bangkok, the average household income is 7,793 baht (NSO n.d.) The second issue is whether the rapid growth of these economies can be thought of as constituting 'indigenous industrialization'. Yoshihara has argued that as the countries depend upon imported skills and technology, and often upon foreign capital and capitalists, South-East Asia represents an example of 'ersatz capitalism' (Yoshihara 1988). It is certainly true that one of the principal challenges facing the ASEAN countries is further to develop both the quality and the quantity of their graduates from higher education.

## Economic stagnation in Vietnam

Myanmar and Indo-China face starkly different conditions – although it should be emphasized that researchers face serious problems of data deficiencies and inaccuracies, and often have to resort to impressionistic statements regarding the progress of the countries in question. Nevertheless economic growth has been disappointing, often stagnant or worse, and the respective governments have arguably been forced due to the deteriorating economic situation to introduce market-based reforms. Vietnam is a case in point. Between 1975, when the North and South were reunified following the end of the Vietnam War and 1980 (the period of the Second Five Year Plan) national income grew by perhaps as little as 0.4 per cent per year – significantly less than the rate of population growth (Kimura 1989: 12). Real incomes actually fell over the period – from an estimated US$199 in 1976 to US$192 in 1979 (Ton That Thien 1983: 711). The major challenge facing the leadership then, and now, is how to implement policies that will reverse the trend. Inevitably, this must first involve an understanding of why the Vietnamese economy should have declined in the first place.

At the risk of simplifying, and it should again be pointed out that accurate information is extremely difficult to come by, the following factors can be highlighted. First, the after-effects of the war were not fully appreciated. There was massive economic dislocation, and its rectification represented at least a medium-term challenge. Further, the Vietnamese invasion of

Cambodia in 1978-9 and its subsequent occupation until 1989 coupled with the border war with China in 1979 did not allow the leadership an opportunity to divert resources from the military to the civil sectors. Second, the US economic blockade and the difficulty of securing assistance from the West, in the form of both trade and aid, forced the Vietnamese to look to the Soviet Union for support. Third, the targets set in the Second (1976-80) and the Third (1981-5) Five Year Plans were grossly over-ambitious, and their priorities unbalanced: heavy industry was stressed at the expense of light industry, and industry in general at the expense of agriculture (see Beresford 1989; Kimura 1989; Tan Teng Lang 1985). However, the most serious constraint to economic growth was the management system: the system of 'bureaucratic centralism and subsidization'. The system was unwieldy, unresponsive to consumer demand, lacking in an effective system of responsibility, and it failed to instil any sense of incentive in those operating within it (Beresford 1989: 203-12). Although minor tinkerings with this system can be traced back to the Sixth Plenum of 1979, by 1985 sections of the leadership were increasingly unrestrained in their criticisms of economic policy. The Planning Minister Vo Van Kiet wrote in 1985:

> Plans are unrealistic and originate from subjective require-
> ments ... planning work has been carried out in accordance
> with a series of criteria imposed by higher echelons ... that are
> not consistent with the various policies and economic incen-
> tives. ... The upper echelons, in the name of the entire cause,
> set very rigorous requirements, but often fail to provide
> sufficient material conditions for the basic units [of produc-
> tion] to carry out their plans.
>
> (quoted in Beresford 1989: 203)

Faced with deteriorating conditions in their own country, and the economic dynamism of ASEAN, the adage that although the Vietnamese may have won the war, they were rapidly losing the peace, seemed to be becoming increasingly trenchant. In response, since the early 1980s and at an increasing pace, the Vietnamese government has instituted a series of economic re-forms as part of a so-called New Economic Policy: agricultural communes and co-operatives have been partially dismantled, private markets in agricultural produce have been permitted, the domestic currency – the dong – has been devalued, foreign investment actively encouraged, wages have been linked to productivity in some industries to increase incentive, and price reforms have been introduced (see Beresford 1989; Thrift and Forbes 1986: 104–17; Kimura 1989; FEER 1985; Stern 1987; Ton That Thien 1983). The process is referred to as *doi moi* – renovation – the Vietnamese equivalent of *perestroika*.[3] Indeed, the IMF has been so impressed by the steps that have been taken that they view Vietnam almost as a role model in structural adjustment (Awanohara 1989). There have been some signs that the reforms are beginning to bite: during the course of the Third Five Year Plan (1981–5) national income grew at an average annual rate of 5 per cent to 6 per cent (estimates vary – Kimura 1989: 12; Wiegersma 1988: 238), while in 1989 Vietnam became the world's third greatest rice exporter shipping 1.4 million tonnes, remarkable considering the near-famine of 1988 (Hiebert 1990). However, even so, the country remains one of the world's poorest, and progress is unsteady. It is notable, although perhaps unsurprising, that it is in the south of the country – capitalist until 1975 – where the reforms have been most successful. In the north, where a command economy on a war footing has been operating for as long as most people can remember, they have been partially effectual at best, and often ineffectual.

## CHANGING ECOLOGY

The economic changes described in the previous section have had inexorable and inevitable consequences for the environment and ecology of South-East Asia (Stott 1990a). In the 1960s, despite the former colonial exploitation of timber resources, much of South-East Asia remained under forest. This forest resource was also a very varied one, ranging from the true lowland evergreen rain forests of Malaysia and Indonesia, which comprise some of the most diverse habitats in the world, through monsoon forests with valuable teak (*Tectona grandis*), dry, seasonal savannah forests, and high montane forests, to more specialist formations, such as the distinctive communities over limestone, peat swamp forests with sago, and coastal mangrove swamps (Whitmore 1984). Many of these formations still have extremely important roles to play in the local economies of South-East Asia, as is well exemplified by the coastal mangroves, which provide quality charcoal for cooking, thatching for houses, as well as *Nypa* fruits, fish and shellfish for food. Even in 1965, it was acknowledged that careful management and development of some of these resources might provide a good basis for sustainable growth.

Today, however, South-East Asia is no longer a forested land. In 1960 Thailand, for example, was still well over 50 per cent forest; yet, by 1990, this figure had dropped to between 10 and 14 per cent, although some government departments and other bodies continue to state publicly that Thailand remains, seemingly in perpetuity, 30 per cent forested (Thailand Development Research Institute 1987). This figure actually refers to *forest land* defined in legal terms; the fact that much of it hardly carries a tree is resolutely overlooked. The above example highlights the extreme difficulty in obtaining satisfactory figures for forest decline in the region, where political sensitivity over the issue, the lack of a clear definition of what is being included under 'forest', and the absence of an agreed forest classification make scientific reliability impossible to achieve. According to the World Resources Institute, the annual rates of deforestation in South-East Asia between the years 1981 and 1985 varied from 0.5 per cent for Indonesia to 2.6 per cent for Thailand, which undoubtedly has one of the worst records in the world. Allen and Barnes (1985) actually give a figure of 3.15 per cent for Thailand, while Grainger in Sutton et al. (1983) has predicted that Asia and the Pacific have only 164 years to complete deforestation.

Quibbling over the regional and global statistics, however, is hardly necessary, as it is now generally agreed that deforestation in South-East Asia has been severe. The causes of forest loss are also many and varied. First, despite attempts to put the ultimate blame largely on agriculturalists and shifting cultivators, logging and straight commercial timber extraction remain a significant factor in the process. Between 1950 and the early 1970s the world consumption of tropical hardwoods probably quadrupled, and most of the new demand has been met from exploiting the forest resources of South-East Asia. In 1983 the countries of ASEAN produced 42.5 cubic metres of logs, 45 per cent of the total production of the tropics (Tesoro 1987). At that time, ASEAN was estimated to have 3,551 saw mills with capacities of more than 500 sq. m. per annum. Moreover, both controlled and uncontrolled timber extraction involve the construction of roads, which then open up formerly protected areas to yet further damage. For example, a programme of 10 per cent timber extraction will tend to leave up to 50 per cent of the forest area damaged. The replacement of the elephant by log-hauling vehicles and the introduction of the one-man chain saw have further exacerbated the process. Much of the felling is also illegal, aided and abetted by corrupt forest officials, an often endemic system of 'cronyism' – a near fine art in the Philippines – and government agencies. If cutting is banned in one location, the loggers tend to move their operations elsewhere, as happened with a recent logging ban in Thailand, which simply exported the problem to Myanmar and Laos, while being only partially effective in Thailand itself.

It cannot be denied, however, that both the controlled and uncontrolled extensification of agriculture have also proved a potent cause of forest loss and degradation, especially where marginal areas are being opened up by shifting cultivators, squatter farmers and opportunistic settlers. Similarly, government-directed pro-grammes, such as the transmigration policy effected in Indonesia (see pp. 83–4), have resulted in the development of forest areas in ecologically sensitive habitats, which are then often badly farmed and managed. Settlement of the uplands by lowland migrants and their consequent failure to respect customary land ownership has further resulted in land grabbing and the alienation of tradi-tional peoples from their principal resources. This often forces the dispossessed to seek new resources and to clear yet more forest, a process well-exemplified among the Negrito hunter-gatherers of the Philippines (Rigg 1991a: 70–8), and one often initiated and exacerbated through government or official action. In 1969 the Presidential Assistant on National Minorities (PANAMIN) attempted a resettlement of the entire Batak population of the island of Palawan; this left the logging interests in Palawan to do as they pleased, so that, according to one recent estimate, out of 780,000 hectares of forest in Palawan, 60 per cent will be lost over the next twenty years (Clad and Vitug 1988).

Mining, the development of hydro-electric power and irrigation dams (Rigg 1991b), other forms of exploitation, such as drilling for oil in Indonesia and the dredging of coastal tin in Malaysia and southern Thailand, coupled with the misuse of fire, a potent ecological force, have also played their part in the destruction of the forests of the region. The impact of war has also been significant. In the late 1960s the military forces of the United States of America are believed to have sprayed some 70 million litres of herbicides (mainly 2, 4, 5, -T) on to the forests of Vietnam, completely sterilizing the coastal mangroves. Many of these ecosystems remain barren even today, despite attempts at replanting. Similarly, war in Cambodia has brought the world's largest cattle, the kouprey (*Bos sauveli*), to the point of extinction. This fine animal, the very symbol of the country, once roamed in great herds through the monsoon and savannah forests of Cambodia, but its population is now decimated through hunting, by desperate villagers and armies seeking much-needed food, through the loss of habitat, and by blowing itself up on mines cruelly laid in the forest.

To some extent, there is at last a reaction to this process of forest destruction, which has proceeded largely unchecked for the last thirty or so years. Increasing evidence that watershed mismanagement is leading to severe soil erosion, to downstream siltation, to the destruction of fisheries and coral reefs, to the undermining of lowland agricultural systems, to local droughts, and to the loss of potentially significant genetic material and breeding animals, such as wild deer, has brought international condemnation down on the heads of countries such as Indonesia, Thailand and the Philippines. Yet, perhaps, more significantly it is now engendering strong local, internal opposition to many development plans, and to corruption in forestry and hunting. The decision in March 1988 of the thirty-nine-member inquiry team, chaired by Deputy Prime Minister Thienchai Sirisamphan, to postpone the building of the proposed Nam Choan Dam in the famous Thung Yai Wildlife Sanctuary in western Thailand, is one example of a recent rather spectacular success achieved by one of the new environmental lobbies (Paisal Sricharatchanya 1988a: 24; Rigg 1991b; Stott 1991). Such lobbies often comprise a mixture of middle-class activists, students, local people, foreign environmentalists and media workers. Significantly, some groups are now striving to discover an ethic of conservation and sound management of natural resources based, not on the values of the North, but on indigenous local values. Particularly noteworthy in this respect are the attempts being made in Thailand to develop a Buddhist philosophy of conservation, both at the local and the national level (Stott 1991). Unfortunately, the environmental pressures are very great and have led recently to some environmentally related murders and suicides.

It must be admitted, however, that the interpretation of the science in all this is not quite as certain as many environmentalists would like. For example, some forest loss has always occurred through fire and flood, disease, wind damage, volcanic action and climatic change, and in many circumstances it is not easy to unravel the human impact from natural changes. From June 1982 to May 1983 the forests of East Kalimantan (eastern Borneo) in Indonesia experienced a major series of fires, which seriously affected some 3 million hectares of lowland tropical rain forest, swamp forest and peat forest (Goldammer and Seibert 1990). One of these local fires travelled some 20 km in two months, while another moved at about 500 m a day. Overall, the fires were one of the most spectacular ecological events witnessed in recent times, and their long-term impact is still under study. On the other hand, their cause is known to have been a complex linking of natural and human factors, the initial fire and its essential sustainability having been triggered by an unusual drought in East Kalimantan, brought about by a climatic phenomenon known as the El Niño Southern Oscillation. Once underway, however, and the fire established, the burns were then spread by poor farming methods, shifting cultivators, and the absence of satisfactory fire education and fire control measures. In fact, throughout South-East Asia, there is a desperate need to understand the ecology of fire more fully, so that proper fire management programmes can be established for the different forest formations (Stott et al. 1990). In some formations, such as the dry, seasonal savannah forests and the monsoon forests, prescribed fire will have to be used as a management tool, and not just simply excluded (Stott 1988).

Similarly, although we certainly know that forest clearance on watersheds can lead to severe soil erosion, it is also apparent that natural geological and geomorphological processes, such as uplift, may also be playing an important role in causing soil erosion and silt production. Again sorting out the relative contributions of the direct human and non-human influences may prove difficult, although the overall process is well attested in South-East Asia, the Red River catchment in Vietnam losing some 1,092 tonnes of soil per sq. km per annum and the delta of the Punagara river in west Java expanding by rates ranging from 0.28 to 0.89 sq km per annum (Donner 1987). Perhaps the point to stress, however, is the need to plan for the most likely scenario, particularly in respect of such major world phenomena as global climatic change and 'global warming'.

In contrast to this wider uncertainty, pollution is now unquestionably a serious problem throughout South-East Asia, both in the air, on the land, and at sea. In cities such as Bangkok and

Metro Manila, National Environmental Board standards for carbon monoxide (CO), total suspended particulates (TSP) and lead (Pb) are frequently exceeded around main city interchanges (Thailand Development Research Institute 1987). Water pollution is, however, even more serious, with some *khlong* (canals) in Bangkok, for example, having a dissolved oxygen count of zero. Throughout the region, many open water bodies are carrying sewage rather than normal water, and these are flowing into major river systems, which, in turn, can become dead in their lower reaches. At sea, the problem is compounded by the fact that South-East Asia straddles the sea routes linking industrial Japan with the oil-rich Middle East. One estimate is that there are some 37,000 supertankers and other forms of shipping passing through the Straits of Malacca annually, with the inevitable disasters and oil spills, such as that from the Japanese registered *Showa Maru* in 1975.

In 1965 South-East Asia was still a land-, forest- and resource-rich region. This is no longer the case, and many of its nation states have wasted their major environmental assets, either for short-term gain or simply through poor and corrupt government, although war has also tended to hinder the satisfactory management of resources. The consequences of all this mismanagement are now being experienced, and a 'greening' of South-East Asia is occuring, but perhaps too slowly and too late. The way forward must include the maintenance of major genetic reserves and wildlife sanctuaries, like Huai Kha Khaeng Wildlife Sanctuary in Thailand, for which World Heritage status should be sought, the development of genuine social forestry and agroforestry programmes, the careful protection and management of watersheds, and the wider and more strict application of environmental laws. For these to succeed, however, there has to be the development of mature political systems, which can curb corruption and exploitation, and plan for the long-term, not simply responding to short-term needs and events. Unfortunately, where the environment is concerned, the 'widespread lawlessness and corruption' described by Fisher in 1966 (see p. 74) remains endemic to South-East Asia. However, the development of a middle-class, who are increasingly sensitive to world criticism, who care for the future of their country, who have a greater vested interest in tourism and recreation, and who are seeking a conservation rationale in traditional values, rather than simply copying those of the North, may be one important factor for change, coupled with the growth of more local pressure groups. Despite the global implications of many of the problems involved, the real answer now has to come from South-East Asians themselves.

One final problem, very much involving South-East Asians, is the changing epidemiology of disease in the region. In 1965 Fisher (1966: 56) wrote that 'the general state of physical health among the peoples of South-East Asia ... presents an extremely sombre picture', and that 'vast expenditures on drugs, equipment and trained personnel, and equally far-reaching changes in many established habits of sanitation and diet will be necessary before the vicious circle of disease, debility, inefficiency, malnutrition and more disease is finally wiped away from the rural areas in which the great majority of people live.' Paradoxically, forest clearance, the spread of agriculture, and improved medicine in some of the countries of the region, especially Malaysia and Singapore, have gone a long way to remedy this dire picture, although diseases such as cholera, typhoid and rabies can still be serious, even in the main cities. Unfortunately, the disease situation is being exacerbated by the development of a suite of new threats, associated with dam development (water-borne diseases), pollution, but above all with the alarming growth of sexually transmitted diseases, such as AIDS. The main cities of the region, especially Bangkok, have become notorious as international sex tourist centres, Bangkok building on its own centuries-old, traditional sex industry as well as the 'rest and recreation' (R&R) programmes of the Vietnam War period. On one estimate, there is now a one in seven chance of becoming HIV-positive through sexual contact with prostitutes in Thailand. Some traditional diseases, such as malaria, have also proved remarkably resistant to eradication, and are flaring up again in new and dangerous forms. The battle with disease is not won, either in the countryside or the city.

## URBANIZATION

### Urbanization in Vietnam

Urbanization in South-East Asia, like economic development has exhibited a similar split between the market and the command economies. Perhaps most extraordinary of all, the cities of Cambodia and in particular the capital Phnom Penh were systematically depopulated – for ideological reasons – during the years of Khmer Rouge government under Pol Pot (Table 4.5). Saigon, the capital of

*Table 4.5* Urbanization in South-East Asia 1965-88

|  | Average annual growth of urban population (%) | |
| --- | --- | --- |
|  | *1965-80* | *1980-8* |
| **ASEAN** | | |
| Brunei | — | — |
| Indonesia | 4.7 | 4.8 |
| Malaysia | 4.5 | 4.9 |
| Philippines | 4.0 | 3.7 |
| Singapore | 1.6 | 1.1 |
| Thailand | 4.6 | 4.7 |
| Weighted average | 4.5 | 4.6 |
| **Indo-China and Myanmar** | | |
| Vietnam | n.a. | 3.9 |
| Laos | 5.3 | 6.1 |
| Cambodia | −0.5 | n.a. |
| Myanmar | 3.2 | 2.3 |

*Source*: World Bank (1990).

South Vietnam until the end of the Vietnam War, and now named Ho Chi Minh City (although many of the inhabitants continue to talk of Saigon), has also experienced a turbulent twenty-five years.

During the Vietnam War, supported with huge quantities of American aid, Saigon had booming – albeit economically unsustainable – informal, black market and service sectors. There were, for example, 56,000 *registered* prostitutes alone (Sheehan 1989: 625). With conditions in the countryside deteriorating as the Vietcong extended their control, there was a migration of scared and displaced rural inhabitants to the cities of the South. As Sheehan writes, 'many of the prostitutes [in Saigon] were farm girls, for another collateral effect of the physical destruction of the countryside was to help fulfil American needs for labour and entertainment' (1989: 625). Da Nang, South Vietnam's second city, grew by over 21 per cent (58,000) between 1967 and 1968 following the Tet offensive, and grew by another 200,000 after the 1972 offensive (Desbarats 1987: 47). Between 1964 and 1974–5 Saigon's population expanded from 2.4 million to 4–4.5 million (Thrift and Forbes 1986: 124). With the final victory of the communist North in 1975, and the reunification of the South and the North, the new government proposed a policy of *de-urbanization* in the South to correct what they felt was an unhealthy level of urbanization, and unhealthy for two reasons: first because the cities did not have the industrial and service (water, housing, electricity) infrastructure to support such large populations without US aid; and second because the bloated cities with their large informal, black market and service sectors were reflections of a despised capitalist past. Some of the 'excess' population drifted back to their villages of its own accord, now that hostilities had ended. Many others – and especially the ethnic Chinese – were harried by the authorities and felt themselves forced to leave. Between 1977 and 1982 709,570 refugees, many of them 'boat people', left Vietnam (Thrift and Forbes 1986: 133–4; and see Thrift 1987). But the government also designated New Economic Zones in the countryside and, in the course of the Second Five Year Plan (1976–80) proposed moving 370,000 people from the now renamed Ho Chi Minh City alone, 1,472,000 in all (Thrift and Forbes 1986: 130; Hill 1984: 390; Desbarats 1987). A target population of only 750,000 was initially set for Ho Chi Minh City, later raised to 1,750,000 (Desbarats 1987: 50).

To end the perceived parasitic role of the southern cities, the authorities attempted to boost urban food production, as well as pursuing the co-operativization of industry and the persecution of the petty bourgeoisie. The production of rice and other cereals in the suburbs of Ho Chi Minh City increased from 95,100 tonnes in 1975, to 200,000 tonnes by 1980 (Thrift and Forbes 1986: 155). In spite of data deficiencies, it seems certain that the service sector in Ho Chi Minh City and elsewhere in south Vietnam has contracted significantly, and that following the de-urbanization of the period between 1975 and 1980 (Ho Chi Minh City had a population of 3.4 million in 1979), the cities of the South have now entered a period of slow, controlled, urban growth.

## Urbanization in ASEAN: Bangkok

Like the countries of Indo-China and Myanmar, policies designed to control the rate of urban growth have also been introduced in some of the countries of ASEAN. The Indonesian government for example declared in 1970 that Jakarta was a 'closed' city to all new migrants. Before they could obtain residence permits for the capital they had to produce evidence that they were gainfully employed, had accommodation, and had sufficient funds to get home again (Yue-man Yeung 1988: 171). But in a market economy there is far less scope to 'control' urbanization than in a command economy. Regional growth pole policies and the promotion of agricultural development can, in theory, discourage migration from the country-side to urban areas, or from secondary cities to the capital, but among the countries of ASEAN such policies have been generally ineffective. Among the countries of ASEAN, the urban population grew at an average annual rate (weighted) of 4.5 per cent between 1965 and 1980, and at 4.6 per cent between 1980 and 1988 (Table 4.5). Over the same periods, population growth was 2.4 per cent and 2.2 per cent respectively.

Such rapid growth has inevitably placed strains upon metro-politan authorities that lack the funds and the man-power to expand services at an equivalent rate (two exceptions to this rule are Singapore and Bandar Seri Begawan, the capital of Brunei Darussalam). The provision of basic human needs to the urban poor of the region is one, important, issue which has received consider-able attention. Rather more nebulous, although none the less also important, is the question of whether the capital primate cities of ASEAN – Jakarta, Kuala Lumpur, Manila, but most clearly Bangkok – are *too* large. In 1988 Bangkok had an official population of 5,670,000 although most analysts believe that the true figure is between 7 million and 10 million. Even taking the official figure, this means that Bangkok is almost twenty-three times larger than Thailand's second city, Chiangmai. But Bangkok is also larger than life in a number of other respects. It has the greatest concentration of universities, hospitals, doctors, industries, financial organizations, telephones and private cars. It generates 45 per cent of the country's wealth, processes 95 per cent of imports and exports, and has an average income over twice that of the national average (Table 4.6). The city lies at the core of the Thai Kingdom's transport network, and houses the key centres of political, military and economic decision-making. Even fashions are dictated, seemingly, by the whim of the city. For an ambitious man or woman it is tantamount to a dictum that they be based in Bangkok. There are two important linked questions that arise from these observations: first, what has caused this level of primacy to arise? And second, is it 'efficient' – in an economic sense? It is also worth pointing out that it was just this degree of primacy or urban bias that led the governments of the communist nations to pursue policies of de-urbanization.

People who call for the control of capital city growth in Thailand and elsewhere believe that the dominance, or super-eminence, of these primate cities is not part of a natural process. In their eyes it is due to policies – both explicit and implicit – which promote capital

*Table 4.6* Indicators of primacy in Bangkok

| | Bangkok | Whole country | Bangkok as % or multiple of whole country |
|---|---|---|---|
| Population (1988) | 5,670,692 | 54,465,056 | 10.4% |
| Economy | | | |
| Gross regional/domestic product (millions baht) | 495,310 | 1,098,366 | 45.1% |
| Average household monthly income (baht, 1986) | 7,427 | 3,710 | 2 times |
| Domestic telex services (Jan–June 1987) | 1,726,758 | 2,456,940 | 70.3% |
| Telephone line capacity | 861,392 | 1,251,102 | 68.9% |
| Health | | | |
| No. of hospital beds | 16,461 | 84,438 | 19.5% |
| No. of physicians | 4,142 | 9,464 | 43.8% |
| No. of dentists | 883 | 1,395 | 63.3% |
| No. of pharmacists | 2,783 | 3,356 | 82.9% |
| Maternal death-rate (per 1,000 population) | 0.04 | 0.35 | — |
| Education | | | |
| No. of state institutes of higher education | 8 | 14 | 57.1% |
| No. of graduates (1985) | 37,858 | 53,492 | 70.8% |
| Transport | | | |
| No. of passenger cars registered (1986) | 593,505 | 753,326 | 78.8% |
| Social | | | |
| No. of divorce licences (1987) | 8,773 | 31,068 | 28.2% |
| Divorce rate (per 1,000) | 1.77 | 0.69 | 2.5 times |
| No. of cinema seats | 99,932 | 399,818 | 25.0% |
| Colour television sets per 100 private households | 56 | 21 | 2.5 times |

city growth at the expense of the countryside. Such policies have contributed to an 'urban bias' in development, or more accurately a 'capital city bias' (Lipton 1977). They can range from the subsidization of urban transport, to tax incentives for industries locating in and around the capital city, to the stifling of agriculture. These policies help to make urban areas, and in particular the capital city, relatively more attractive than other parts of the country. In Thailand, hundreds of thousands of migrants (the exact figure is not known) flow from the countryside into Bangkok each year in the search for work and a better life, swelling the population and straining the infrastructure of the city. The other side of the urban bias/primacy coin is rural stagnation. If cities are advantaged to the extent that wages are considerably higher than those in rural areas, then farmers will reduce agricultural investment and in many cases will move to urban areas. The resulting fall in agricultural production can have a serious effect on food self-sufficiency and livelihoods in the countryside. In Peninsular Malaysia for example, the rural to urban shift in population has directly led to an estimated 810,000 hectares of land in the smallholder sector being left idle (Courtenay 1987). Interestingly, in Malaysia's case, this figure would be even higher if it were not for the influx of Thai workers from across the border, attracted by the relatively high agricultural wages being offered.

It is argued that the excessive growth of Bangkok and other capital cities leads to a situation in which the city is so large that the costs of congestion, pollution, land and labour actually make it less efficient than other towns – so-called diseconomies of scale. This is all too believable when it takes a visitor two hours to travel a few kilometres in Bangkok. The problem, however, is that there is very little evidence to support the contention that diseconomies of scale are beginning to assert themselves. Even given the policies noted above which favour capital cities, Bangkok is still the most economically sensible place for most companies to locate. Should congestion become so serious as to undermine profitability then companies will, of their own accord, move elsewhere. In other words, there will be a market-based adjustment (e.g. Jones 1988: 142).

## POLITICAL CHANGE

### Confrontation and conciliation in South-East Asia

In the introduction to this chapter, it was noted that since 1965 there have emerged three 'South-East Asias': the market economies of ASEAN, the command economies of Indo-China, and Myanmar. The region has effectively been polarized into distinct and competing sub-regional blocs. It is this polarization which has dominated regional affairs over the last twenty-five years.

In 1967, as US involvement in Indo-China was reaching its height, Thailand, Malaysia, Singapore, Indonesia and the Philippines signed the Bangkok Declaration and founded the Association of South-East Asian Nations or ASEAN. With the victory of communist forces in Vietnam, Laos and Cambodia in 1975, and the complete withdrawal of US forces from mainland South-East Asia, the Association was forced to reassess its relations with the rest of the region, and in particular with the Socialist Republic of Vietnam. Although they were worried about developments in Indo-China, ASEAN decided to pursue relations on the basis of the principles of peaceful coexistence. But Hanoi rejected these overtures of friendship, viewing the grouping as essentially a pro-Western and anti-communist alliance. The political, and by implication also economic, polarization of the region was finally assured when the People's Army of Vietnam invaded Cambodia on Christmas Day 1978. As Weatherbee has noted:

> No matter how complex the factors may have been in Hanoi's decision to invade, it was a dramatic demonstration to a worried ASEAN of the willingness of its potential adversary to use force in pursuit of its external political objectives. The perception was that the first South-East Asian 'domino' had fallen to aggressive Vietnamese expansionism.
>
> (Weatherbee 1989: 189)

The Vietnamese invasion and subsequent occupation of Cambodia galvanized ASEAN. Fielding the largest army in South-East Asia and with over 200,000 troops in Cambodia, many close to the border with Thailand, Vietnam was perceived to be a serious threat to ASEAN's integrity. From 1979–89, ASEAN has adopted a twofold strategy towards the Vietnamese occupation of Cambodia. The member states attempted to isolate Vietnam economically, diplomatically and politically, and at the same time, assist in pressurizing the country militarily. The fact that the Vietnamese-backed government in Phnom Penh is still not recognized by the UN shows how successful ASEAN have been as regards the first of these objectives.[4] In addition Vietnam and Cambodia have been denied membership of the World Bank, the Asian Development Bank and the International Monetary Fund and of virtually all non-humanitarian aid and assistance from Western countries. This has forced them to look to the Soviet Union and the Eastern Bloc instead. The second element of ASEAN's strategy has been achieved through encouraging support for the forces of the three rebel factions that comprise the exiled Coalition Government of Democratic Kampuchea.

108

As Leifer has written:

The conflict over Kampuchea has produced within Southeast Asia an unprecedented condition of political polarization. Although such a condition was latent when revolutionary communism took hold in Indo-China in 1975, it was not fully realized until Vietnam, as a matter of strategic necessity, invaded Kampuchea to incorporate it within a structure of political conformity.

(Leifer 1986: 7)

Interestingly, this perspective offered by Leifer has begun to break down, coinciding – and not accidentally – with similar developments in Eastern Europe and the former Soviet Union. At the end of September 1989, Vietnam withdrew most of its forces from Cambodia (Tasker and Hiebert 1989) thus effectively removing the single most significant obstacle to rapprochement between ASEAN and the countries of Indo-China. Thailand, the front-line state in the standoff between the 'East' and the 'West' (for the confrontation in South-East Asia was a reflection of the East–West conflict), significantly softened its stance towards Vietnam as well as Cambodia and Laos during the course of 1989 and into 1990.

But, not only has there been a degree of political convergence, this has also been reinforced by the economic convergence discussed above. Vietnam, Laos and Cambodia, and even Myanmar to a degree, have instituted significant economic reforms. In a speech he gave in the latter half of 1988, the newly elected Prime Minister Chatichai Choonhavan said that he hoped Thailand would assist in turning Indo-China 'from a battlefield into a market place' (Amnuay Viravan 1989: 172). To this end trade restrictions are being removed and business links actively encouraged. Trade between the town of Koh Kong in south-western Cambodia and the Thai province of Trat for example is flourishing, with an estimated US$5.8 million worth of goods passing between the town and Thailand each month (Tasker 1989: 36). In addition, traders in Aranyaprathet, also on the border with Cambodia, are hopeful that the railway line to Battambang and Phnom Penh may be reopened thereby encouraging a further growth in trade. Analysts feel that Thailand and Indo-China are economically complimentary. Thailand can provide the level of technology that Vietnam, Laos and Cambodia need as they strive to modernize their economies, while they in their turn can provide Thailand with the raw materials necessary to fuel its development (Paisal Sricharatchanya 1988b: 88–9). The same argument of economic complimentarity could likewise be applied to the other countries of ASEAN whose economies might also benefit from closer links with Indo-China (Pike 1984).

## REGIONALISM IN SOUTH-EAST ASIA

### Indo-China

The Vietnamese government has long regarded the three countries of Indo-China as 'strategically interdependent', with Vietnam's own security being contingent upon a friendly Laos and Cambodia (Sukhumbhand Paribatra 1987:141–2; Pike 1986: 245).[5] As a result, the leaders of Vietnam found it difficult to accept Laos and Cambodia as neutral buffers if that meant 'having governments in Vientiane and Phnom Penh that are weak, divided and vulnerable to penetration by powers besides Vietnam' (Turley 1989: 185). These fears caused Vietnam to become increasingly dependent on the support of the Soviet Union, and to attempt to forge a political and economic relationship with the other countries of Indo-China.

In June 1978 Vietnam became a full member of the Council for Mutual Economic Assistance (COMECON), and in November of the same year signed a Treaty of Friendship and Co-operation with the Soviet Union (Weatherbee 1985: xii). Although relations with some non-aligned countries – such as India – remained cordial, during the course of the 1970s and 1980s Vietnam increasingly allied itself with the countries of the Eastern bloc and in particular with the Soviet Union. This relationship led to Vietnam becoming economically, militarily and politically dependent on the Soviet Union. By the late 1980s the Soviet Union was providing an estimated US$1 billion–1.5 billion in aid each year, was Vietnam's major trading partner, was supplying food in times of shortage, and was arming and equipping the Vietnamese armed forces (see Finkelstein 1987; Chang Pao-min 1987). The relationship was not entirely one-sided however. The Soviet Union in its turn won access to the valuable abandoned US military facilities at Cam Ranh Bay on the coast of southern Vietnam, and gained a useful client state in South-East Asia.

At the same time, Vietnam tried to extend and deepen its influence over the other countries of Indo-China. Road links between Vietnam and Laos and Cambodia were substantially improved, an oil pipeline constructed between Vinh in Vietnam to Route 13 in Laos, assistance provided to foster increased food production, and there is even evidence of Vietnamese settlement in Laos and Cambodia (so-called Vietnamization) (Turley 1989: 178; Stuart-Fox 1986: 171–83).[6] Politically, the three countries also moved towards greater unity. In July 1977 a Treaty of Friendship and Co-operation was signed between Laos and Vietnam, and in February 1979 one signed between Cambodia and Vietnam (Sukhumbhand Paribatra 1987: 142). Vietnamese military advisers were attached to the Lao army, and cadres from Vietnam also assisted in the political training of Lao officers (Stuart-Fox 1986: 174). Military links between Cambodia and Vietnam became even closer. Vietnamese commentators no longer talked of an Indo-Chinese Federation, but instead used the term 'confederation'. Whatever the semantics of the links might be however, it is clear that Vietnam was keen to ensure that neither Laos nor Cambodia developed close relations with other countries, and that Vietnam remained the dominant partner in the relationship (Pike 1986: 246). To this end, close ties were maintained by Vietnam with the communist parties of Laos and Cambodia, and in particular with their leaders. Laotians have been reported as saying that the affection shared between themselves and the Vietnamese is 'deeper than the water of the Mekong river', while the Vietnamese talk of a relationship that is 'closer than lips and teeth' (Stuart-Fox 1986: 180). The links between the three countries of Indo-China were finally formally endorsed at the first summit meeting of the leaders of Vietnam, Laos and Cambodia held in Vientiane at the end of February 1983 at which they signed a joint statement creating an Indo-Chinese grouping of countries (*Asia Yearbook* 1984: 284).

## The Association of South-East Asian Nations (ASEAN)

ASEAN has already been mentioned. Formed in 1967 in response to shared fears about the security of the region, the initial eight years of the Association's existence represented a period of 'growing together'. Very little tangible was achieved, although it might be argued that the mere fact that the grouping survived was achievement enough. With the victory of communist forces in Indo-China in 1975, ASEAN was forced to reassess its relations with the rest of the region. The Bali Summit of 1976 was the first attempt to accelerate and intensify the degree of co-operation between the member states. The Summit resulted in a commitment to greater political co-operation, and also set a framework through which greater economic co-operation could be achieved. On the latter point, it was agreed that ASEAN Industrial Projects (AIPs) should be established and that preferential tariffs should be exchanged (PTAs – Preferential Trading Arrangements). When Vietnam invaded Cambodia in 1978–9 the need for group action against what was perceived to be an expansionist and aggressive Vietnam (with the world's fourth largest army) became starker still. The solidarity that ASEAN has shown in response to the Vietnamese occupation of Cambodia can be highlighted as its single greatest success. That said, as the situation in Cambodia and Indo-China in general is gradually resolved, so the focus of Asean's greatest success will correspondingly disappear. Without the galvanizing effect of an external threat, there may be less to hold the grouping together. However, ASEAN was not established as a security grouping but, primarily, as one to promote economic co-operation. On this front, it has been far less successful.

If intra-ASEAN trade as a proportion of total trade can be taken as a broad measure of the degree to which co-operation has been engendered then progress has been minimal. Although there have been plethora of directives and agreements – PTAs, AIPs, AICs (ASEAN Complementation Scheme), AIJVs (ASEAN Industrial Joint Ventures) – the proportion of intra-regional trade has declined from 18.3 per cent in 1967, to 16.9 per cent in 1988 (IMF 1988). Why should there have been such unimpressive progress when ASEAN is often perceived as the most successful regional association in the developing world (e.g. Hill n.d.: 82)? Very briefly, four reasons can be highlighted. First, it is often stressed that the countries of ASEAN are competitive rather than complimentary. They tend to specialize in the same exports – primary products and intermediate manufactures. Second, the operation of ASEAN itself is not conducive to forcing through difficult decisions: it operates on the basis of consensus. Third, and related to this second point, only rarely have the leaders of the member states been willing to make short-term national sacrifices in order to reap long-term regional benefits. And fourth, the very success of the ASEAN economies (with the exception of the Philippines) has made it less urgent that regional co-operation be fostered. But should, for example, global protectionism increase, then it would be likely that the member states might look more assiduously at the possibilities for greater regional co-operation.

There is one country of South-East Asia whose role and place in the region has not yet been discussed: Myanmar. As noted in the introduction to this chapter, Myanmar represents a third South-East Asia. Since 1962, when General Ne Win assumed the leadership of Myanmar (then Burma) for the second time and embarked on his Burmese Way to Socialism, the country has been virtually cut off from the outside world and has followed an autarkic and idiosyncratic development strategy. Elements of the strategy mirror those of the countries of Indo-China: the nationalization of all foreign and privately owned firms and banks (15,000 in all), and state control over the production, distribution, import and export of all major commodities (Steinberg 1982: 77–9). In short, state ownership of the basic means of production. However, in spite of the superficial similarities between the economic policies pursued by Myanmar and the countries of Indo-China, there has been little to unite them politically. On the whole, Myanmar's foreign relations have been limited and restrained, and the country has refrained from forming alignments with any other state. Since 1962 Myanmar has shifted from neutralism to extreme isolationism and back towards a modified form of neutralism (Steinberg 1986: 256–9).

The ASEAN state with the closest links with Myanmar is Thailand. This is largely for geographical reasons, and a number of joint ventures and other agreements – aimed, for example, at the exploitation of Burma's marine and timber resources – have recently been established or concluded. In very simple terms, Thailand finds that it is short of the natural resources to supply its own timber and fishing industries (having depleted its own), while Myanmar is short of foreign exchange. As many as forty timber concessions along the border between Myanmar and Thailand were awarded to Thai companies between 1988 and 1989, and levels of trade have increased accordingly (McDonald 1990). Nevertheless, Myanmar remains very much the maverick country of South-East Asia, and until General Ne Win dies, or is *de facto* replaced, it is unlikely that this state of affairs will change.

## TOWARDS A NEW, OLDER GEOGRAPHY

In 1971 and 1979 Tate produced two volumes of his substantial *The Making of Modern South-East Asia*, the first sub-titled *The*

*European Conquest*, the second *The Western Impact*. Both sub-titles reflect clearly the fact that the history and geography of South-East Asia were, until very recently, completely dominated by a powerful European paradigm, which emphasized the significance of the colonial legacy. In fact, Tate explicitly stated that the theme of his second volume was above all the economic 'imprint of the West' (1979: 1), of the new industrial societies which arose out of 'the phenomenon of the Industrial Revolution'. Similarly, Fisher's (1964) *Geography* is essentially a colonial and post-colonial study of the region. He writes, in his chapter on 'The Legacy of the West',

> however much the era of Western rule may be resented, the economic and social conditions which it has left provide the only realistic basis on which to start building now, even if the new architects aspire to erect new structures of a very different kind in the years that lie ahead.
>
> (Fisher 1966: 199)

But what both these writers perhaps failed to see is the possibility that the 'new' structures might actually lie on foundations much older than the colonial experience which they documented so faithfully and thoroughly. Beneath the veneer of the West, there remained, often little altered, age-old human–land and human–water relationships, which had characterized South-East Asia from before the first century AD, and which represent the real, essential 'geography' of the region. Often our interpretation of these relationships has itself been over-strongly influenced by current Western intellectual traditions, so that the great civilization of Angkor (ninth to fifteenth centuries AD), for example, was presented as a prime example of a 'hydraulic society', in the sense developed by Wittfogel (1957), when the actual evidence was clearly pointing to another, more South-East Asian interpretation (e.g. Van Liere 1982; Stott 1990b). This then is one major aspect of our present 'Geography of Ignorance' concerning South-East Asia. There is a desperate need to understand far more fully the indigenous evolution of human–land and, especially for South-East Asia – the realm of the Naga, with its rivers, lakes and central sea – of human–water relationships, so that we can interpret more profoundly the 'new' structures which have developed on a much older geography.

To do this, however, we will need many more indigenous South-East Asian geographers, a region in which, with the exception of

Singapore, the subject is far from strongly represented in universities and colleges, although it is good to note that HRH Princess Maha Chakri Sirindhorn of Thailand is setting a fine example which others should strive to follow. It is thus vital that an intellectual tradition of geography becomes established more widely in the region as soon as possible. Then, the celebratory volumes of the future will above all be able to record the achievements of South-East Asian geographers, not just those of the geographers of South-East Asia.

## NOTES

1 Rather more contentiously, among the shifting cultivating hill tribes of the northern region, who often live far from towns and health clinics, the programme enthusiastically promoted the use of Depro Medroxy Projesterone Acetate (DMPA), a long-term injectable contraceptive. This led to a dramatic increase in the use of contraceptives, and a corresponding decrease in birth and population growth rates amongst the tribal peoples. In 1975 there were 12,376 new family planning acceptors using DMPA in the northern region. By 1980 this had increased to 39,090, and by 1988 to 140,728.

2 The province of Irian Jaya has an average population density of three persons per sq km (Hugo et al. 1987).

3 In Laos a similar process of reform has also been underway – *chin thanakaan mai* or 'new thinking'.

4 Even more remarkable when it is remembered that the Cambodian seat in the UN was occupied by the ASEAN-sponsored Coalition Government of Democratic Kampuchea (CGDK), which counts among its members the universally vilified Khmer Rouge.

5 The Vietnamese general Vo Nguyen Giap wrote in 1950 while fighting against the French: 'Indochina is a single strategic unit, a single battlefield, and here we have the mission of helping the movement to liberate all Indochina' (quoted in Sukhumbhand Paribatra 1987: 141).

6 The numbers of Vietnamese settlers in Laos and Cambodia are the subject of dispute. In the case of Cambodia, the CGDK claim that over 1 million Vietnamese have been settled in Cambodia, while the Vietnamese-backed government in Phnom Penh state that the true figure is only 80,000. Another estimate puts the numbers of Vietnamese settlers at between 400,000 and 450,000 (in a population of some 6.7 million) (Hiebert 1989: 38).

## REFERENCES

Alexander, J. and Alexander, P. (1982) 'Shared poverty as ideology: agrarian relationships in colonial Java', *Man (N.S)* 17: 597–619.

Allen, J.C. and Barnes, D.F. (1985) 'The causes of deforestation in developing countries', *Annals of the Association of American Geographers* 75(20): 163–84.

Amnuay, Viravan (1989) 'Southeast Asia: turning a battlefield into a market place', *Bangkok Bank Monthly Review* 30(4): 172–5.

Arndt, H.W. (1983) 'Transmigration: achievements, problems, prospects', *Bulletin of Indonesian Economic Studies* 19(3): 50–73.

Arndt, H.W. (1984) 'Little land many people ...', *Far Eastern Economic Review* 1 November: 40.

*Asia Yearbook* (1984) *Asia Yearbook 1984*, Hong Kong: Review Publishing Company Limited.

*Asia Yearbook* (1990) *Asia Yearbook 1990*, Hong Kong: Review Publishing Company Limited.

Awanohara, S. (1989) 'Fiscal interdiction', *Far Eastern Economic Review* 28 September: 22–3.

Barker, R. (1985) 'The Philippine Rice Program – lessons for agricultural development', in Salas, R.M. (ed.) *More than the Grains: Participatory Management in the Rice Sufficiency Programme 1967–1969*, Tokyo: Simul Press.

Barker, R. Herdt, R.W. and Rose, B. (1985) *The Rice Economy of Asia*, Washington DC: Resources for the Future.

Bauer, J. and Mason, A. (1990) 'Fertile future', *Far Eastern Economic Review* 17 May: 44–5.

Beresford, M. (1989) *National Unification and Economic Development in Vietnam*, London: Macmillan.

Booth, A. (1985) 'Accommodating a growing population in Javanese agriculture', *Bulletin of Indonesian Economic Studies* 21(2): 115–45.

Caufield, C. (1985) *In the Rainforest*, London: Heinemann.

Cederroth, S. and Gerdin, I. (1986) 'Cultivating poverty: the case of the Green Revolution in Lombok', in Norlund, I., Cederroth, S. and Gerdin, I. (eds) *Rice Societies: Asian Problems and Prospects*, Scandinavian Institute of Asian Studies, London: Curzon Press.

Chang Pao-min (1987) 'Kampuchean conflict: the continuing stalemate', *Asian Survey* 27(7): 748–63.

Clad, J. (1989) *Behind the Myth: Business, Money and Power in Southeast Asia*, London: Unwin Hyman.

Clad, J. and Vitug, M.D. (1988) 'The politics of plunder', *Far Eastern Economic Review* 24 November: 48–50.

Colchester, M. (1986) 'The struggle for land: tribal peoples in the face of the transmigration programme', *The Ecologist* 16(2/3): 99–110.

Collier, W.L. (1981) 'Agricultural evolution in Java', in Hansen, G.E. (ed.) *Agriculture and Rural Development in Indonesia*, Boulder, Col.: Westview Press.

Collier, W.L., Gunawan Wiradi, and Soentoro (1973) 'Recent changes in rice harvesting methods: some serious social implications', *Bulletin of Indonesian Economic Studies* 9(2): 36–45.

Collier, W.L., Soentoro, Gunawan Wiradi and Makali (1974) 'Agricultural technology and institutional change in Java', *Food Research Institute Studies* 13(2): 169–94.

Courtenay, P.P. (1987) 'Letter from Malacca', *Far Eastern Economic Review* 10 September: 102.

Dalrymple, D.G. (1986) *Development and Spread of High-Yielding Rice Varieties in Developing Countries*, Washington DC: Agency for International Development.

Desbarats, J. (1987) 'Population redistribution in the Socialist Republic of Vietnam', *Population and Development Review* 13(1): 43–76.

Donner, W. (1987) *Land Use and Environment in Indonesia*, London: C. Hurst.

*The Economist* (1989) 'The fifth tiger', 28 October: 15.

Feder, E. (1983) *Perverse Development*, Quezon City: Foundation for Nationalist Studies.

FEER (1985) 'Hanoi's bitter victory', *Far Eastern Economic Review* 2 May: 30–40.

FEER (1989) 'Malaysia: do you sincerely want to be a NIC?', *Far Eastern Economic Review* 7 September: 96–100.

Finkelstein, D.M. (1987) 'Vietnam: a revolution in crisis', *Asian Survey* 27(9): 973–90.

Fisher, C.A. (1964; 2nd edn 1966) *South-East Asia: A Social, Economic and Political Geography*, London: Methuen.

Furnivall, J.S. (1948) *Colonial Policy and Practice: A Comparative Study of Burma and Netherlands India*, New York: New York University Press.

Furnivall, J.S. (1980) 'Plural societies', in Evers, H.-D. (ed.) *Sociology of South-East Asia: Readings on Social Change and Development*, Kuala Lumpur: Oxford University Press.

Goldammer, J.G., and Seibert, B. (1990) 'The impact of droughts and forest fires on tropical lowland rain forest of East Kalimantan' in Goldammer, J.G. (ed.) *Fire in the Tropical Biota*, Ecological Studies 84, Berlin: Springer.

Government of Malaysia (1971) *Second Malaysia Plan 1971–75*, Kuala Lumpur: Government Press.

Hardjono, J. (1986) 'Transmigration: looking to the future', *Bulletin of Indonesian Economic Studies* 22(2): 28–53.

Hardjono, J. (1988) 'The Indonesian transmigration program in historical perspective', *International Migration* 26(4): 427–39.

Hayami, Y. (1984) 'Assessment of the Green Revolution', in Eicher, C.K. and Staatz, J.M. (eds) *Agricultural Development in the Third World*, Baltimore, Md: Johns Hopkins University Press.

Herdt, R.W. (1987) 'A retrospective view of technological and other changes in Philippine rice farming, 1965–1982', *Economic Development and Cultural Change* 35(2): 329–49.

Hiebert, M. (1989) 'No love lost', *Far Eastern Economic Review* 12 October: 38.

Hiebert, M. (1990) 'The tilling fields', *Far Eastern Economic Review* 10 May: 32–4.

Hill, H. (n.d.) 'Challenges in Asean economic cooperation: an outsider's perspective', in Noordin Sopiee, Chew Lay See and Lim Siang Jin (eds) *Asean at the Crossroads: Obstacles, Options and Opportunities in Economic Co-operation*, Kuala Lumpur: Institute of Strategic and International Studies.

Hill, R.D. (1984) 'Aspects of land development in Vietnam', *Contemporary Southeast Asia* 5(4): 389–402.

Hugo, G.J., Hull, T.H., Hull, V.J. and Jones, G.W. (1987) *The Demographic Dimension in Indonesian Development*, Singapore: Oxford University Press.

Hüsken, F. (1979) 'Landlords, sharecroppers and agricultural labourers: changing labour relations in rural Java', *Journal of Contemporary Asia* 9(2): 140–51.

IMF (1989) *Directory of Trade Statistics Yearbook, 1989*, Washington DC: International Monetary Fund.

Jones, G.W. (1988) 'Urbanization trends in Southeast Asia: some issues for policy', *Journal of Southeast Asian Studies* 19(1): 137–54.

Kimura, T. (1989) *The Vietnamese Economy 1975–1986: Reforms and International Relations*, Tokyo: Institute of Developing Economies.

Krannich, C.R. and Krannich, R.L. (1980) 'Family planning policy and community-based innovations in Thailand', *Asian Survey* 20(10): 1,023–37.

Leifer, M. (1986) 'Obstacles to peace in Southeast Asia', in Hiroshi Matsumoto and Noordin Sopiee (eds) *Into the Pacific Era: Southeast Asia and its Place in the Pacific*. Proceedings of the Global Community Forum '84 Malaysia, Kuala Lumpur: Institute of Strategic and International Studies (ISIS).

Lipton, M. (1977) *Why Poor People Stay Poor: A Study of Urban Bias in World Development*, London: Temple Smith.

Lipton, M. with Longhurst, R. (1989) *New Seeds and Poor People*, London: Unwin Hyman.

McDonald, H. (1990) 'Partners in plunder', *Far Eastern Economic Review*, 22 February: 16–18.

MOAC (1989) *Agricultural Statistics of Thailand, Crop Year 1988/89*, Agricultural Statistics no. 414, Bangkok: Ministry of Agriculture and Co-operatives.

NESDB (n.d.) *The Sixth National Economic and Social Development Plan 1987–1991*, Bangkok: National Economic and Social Development Board.

*Newsweek* (1988) 'Asia's emerging superstar', *Newsweek*, 27 June: 6–12.

NSO (n.d.) *Statistical Handbook of Thailand 1989*, Bangkok: Office of the Prime Minister.

Otten, M. (1986) *Transmigrasi: Myths and Realities, Indonesian Resettlement Policy, 1965–1985*, International Work Group for Indigenous Affairs document no. 57, Copenhagen: IEGIA.

Paisal Sricharatchanya (1988a) 'Politics of power', *Far Eastern Economic Review* 31 March: 24.

Paisal Sricharatchanya (1988b) 'First in the door', *Far Eastern Economic Review* 3 November: 88–9.

Pearse, A. (1980) *Seeds of Plenty, Seeds of Want: Social and Economic Implications of the Green Revolution*, Oxford: Clarendon Press.

Penporn Tirasawat (1984) 'Thailand', in Schubnell, H. (ed.) *Population Policies in Asian Countries: Contemporary Targets, Measures and Effects*,

Hong Kong: the Drager Foundation and the Centre of Asian Studies, University of Hong Kong.

Pike, D. (1984) 'Vietnam and ASEAN: the potential for economic intercourse', in Jackson, K.D. and Soesastro, M.H. (eds) *ASEAN Security and Economic Development*, Research Papers and Policy Studies 11, Berkeley, Calif.: Institute of East Asian Studies, University of California.

Pike, D. (1986) 'Vietnam and its neighbors: internal influences on external relations', in Jackson, K.D., Sukhumbhand Paribatra and Djiwandono, J.S. (eds) *ASEAN in Regional and Global Context*, Research Papers and Policy Studies 18, Berkeley, Calif.: Institute of East Asian Studies, University of California.

Rigg, J.D. (1989) 'The new rice technology and agrarian change: guilt by association?', *Progress in Human Geography* 13(3): 374–99.

Rigg, J.D. (1990) 'The Green Revolution: 25 years on', *Geography Review* 4(1): 32–4.

Rigg, J.D. (1991a) *Southeast Asia: A Region in Transition*, London: Unwin Hyman.

Rigg, J.D. (1991b) 'Thailand's Nam Choan Dam project: a case study in the "greening" of South-East Asia', *Global Ecology and Biogeography Letters* 1(2): 42–54.

Rodan, G. (1989) *The Political Economy of Singapore's Industrialization: National State and International Capital*, London: Macmillan.

Rosenfield, A., Bennett, A., Somsak Varakamin and Lauro, D. (1982) 'Thailand's family planning program: an Asian success story', *International Family Planning Perspectives* 8 (June): 43–50.

Schweizer, T. (1987) 'Agrarian transformation? Rice production in a Javanese village', *Bulletin of Indonesian Economic Studies* 23(2): 38–70.

Scott, J.C. (1985) *Weapons of the Weak: Everyday Forms of Peasant Resistance*, Bellhaven, Conn: Yale University Press.

Sheehan, N. (1989) *A Bright Shining Lie*, London: Jonathan Cape.

Steinberg, D.I. (1982) *Burma: A Socialist Nation of Southeast Asia*, Boulder, Col.: Westview Press.

Steinberg, D.I. (1986) 'Burmese domestic politics and foreign policy toward ASEAN', in Jackson, K.D., Sukhumbhand Paribatra and Djiwandono, J.S. (eds) *ASEAN in Regional and Global Context*, Research Papers and Policy Studies 18, Berkeley, Calif.: Institute of East Asian Studies, University of California.

Stern, L.M. (1987) 'The scramble towards revitalization: the Vietnamese Communist Party and the Economic Reform Program', *Asian Survey* 27(4): 477–93.

Stott, P. (1988) 'The forest as phoenix: towards a biogeography of fire in mainland South East Asia', *Geographical Journal* 154(3): 337–50.

Stott, P. (1990a) 'Asia and Pacific ecology', in Taylor, R.H. (ed.) *Handbooks to the Modern World: Asia and the Pacific*, New York and Oxford: Facts on File.

Stott, P. (1990b) 'Angkor: shifting the hydraulic paradigm', unpublished paper presented to the seminar series *The Gift of Water*, Centre of South-East Asian Studies, School of Oriental and Asian Studies, London.

Stott, P. (1991) 'Mu'ang and pa: élite views of nature in a changing Thailand', in Chitakasem, M. and Turton, A. (eds) *Thai Constructions of Knowledge*, London: SOAS.

Stott, P., Goldammer, J.G. and Werner, W.L. (1990) 'The role of fire in the tropical lowland deciduous forests of Asia', in Goldammer, J. (ed.) *Fire in the Tropical Biota: Ecosystem Processes and Global Challenges*, Ecological Studies 84, Berlin: Springer.

Strauch, J. (1981) *Chinese Village Politics in the Malaysian State*, Cambridge, Mass.: Harvard University Press.

Stuart-Fox, M. (1986) *Laos: Politics, Economics and Society*, London: Frances Pinter.

Sukhumbhand Paribatra (1987) 'The challenge of co-existence: ASEAN's relations with Vietnam in the 1990s', *Contemporary Southeast Asia* 9(2): 140–56.

Sutton, S.L., Whitmore, T.C. and Chadwick, A.C. (eds) (1983) *Tropical Rain Forest: Ecology and Management*, Oxford: Blackwell Scientific.

Tan Teng Lang (1985) *Economic Debates in Vietnam: Issues and Problems in Reconstruction and Development (1975–1984)*, Research notes and discussion papers no. 55, Singapore: Institute of Southeast Asian Studies.

Tasker, R. (1989) 'Peaceful coexistence', *Far Eastern Economic Review* 12 October: 36.

Tasker, R. and Hiebert, M. (1989) 'A test of arms', *Far Eastern Economic Review* 28 September: 20–1.

Tate, D.J.M. (1971) *The Making of Modern South-East Asia, Vol. I: The European Conquest*, Kuala Lumpur: Oxford University Press.

Tate, D.J.M. (1979) *The Making of Modern South-East Asia, Vol. II: The Western Impact*, Kuala Lumpur: Oxford University Press.

Tesoro, F. (1987) 'Tropical timbers for construction in the ASEAN countries', *Unasylva* 156, 39(2): 20–6.

Thailand Development Research Institute (1987) *Thailand: Natural Resources Profile*, Bangkok: TDRI.

Thrift, N. (1987) 'Vietnam: geography of a socialist siege economy', *Geography* 72(4): 340–4.

Thrift, N. and Forbes, D. (1986) *The Price of War: Urbanization in Vietnam 1954–1985*, London: Allen & Unwin.

Ton That Thien (1983) 'Vietnam's new economic policy', *Pacific Affairs* 56(4): 691–712.

Turley, W.S. (1989) 'Vietnam's strategy for Indochina and security in Southeast Asia', in Young Whan Kihl and Grinter, L.E. (eds) *Security, Strategy and Policy Responses in the Pacific Rim*, Boulder, Col.: Lynne Rienner Publishers.

Van Liere, W.J. (1982) 'Was Angkor a hydraulic society?', *Ruam Botkwan Prawatisat* 4: 36–48.

Weatherbee, D.E. (1985) 'Preface', in Weatherbee, D.E. (ed.) *Southeast Asia Divided: The ASEAN–Indochina Crisis*, Boulder, Col.: Westview Press.

Weatherbee, D.E. (1989) 'ASEAN defense programs: military patterns of national and regional resilience', in Young Whan Khil and Grinter, L.E. (eds) *Security, Strategy and Policy Responses in the Pacific Rim*, Boulder, Col: Lynne Rienner Publishers.

Whitmore, T.C. (1984) *Tropical Rain Forests of the Far East*, 2nd edn, Oxford: Clarendon Press.

Wiegersma, N. (1988) *Vietnam: Peasant Land, Peasant Revolution*, London: Macmillan.

Wittfogel, K.A. (1957) *Oriental Despotism*, New Haven, Conn.: Yale University Press.

World Bank (1988) *Indonesia: The Transmigration Program in Perspective*, Washington DC: World Bank.

World Bank (1990) *World Development Report 1990*, New York: Oxford University Press.

Yoshihara, K. (1988) *The Rise of Ersatz Capitalism in South-East Asia*, Singapore: Oxford University Press.

Yue-man Yeung (1988) 'Great cities of Eastern Asia', in Dogan, M. and Kasarda, J.D. (eds) *The Metropolis Era, Vol. 1: A World of Giant Cities*, Newbury Park, Calif.: Sage.

# 5

# THE CHANGING GEOGRAPHY OF THE PEOPLE'S REPUBLIC OF CHINA

*Michael Freeberne*

## INTRODUCTION

China represents certainly the oldest and arguably the greatest major civilization on earth. This is a fact which has sustained within China the belief in an inherent superiority, attendant if intermittent xenophobia, and insistence upon self-sufficiency in political, technical and economic spheres.

The proclamation of the People's Republic of China by Mao Zedong from a rostrum in Tiananmen Square on 1 October 1949 apparently constituted a radical break with the past. Yet important historical continuities persist. In the last century imperial China suffered many indignities at the hands of Western powers. As the Empire gave way to a Republic in 1912, China was to experience further humiliation during the course of two world wars, compounded by civil war and domestic warlordism. In particular, China was invaded by the Japanese, and not even the United Front could resolve the differences between the Guomindang and the communists. A century and a decade of more or less unremitting strife from 1840 to 1949 sorely sapped the energies of the nation, which made the early successes of the New China the more remarkable.

The epic Long March saw Mao Zedong established in the summer cool and winter warmth of the Ya'an caves. After the communist victory in 1949, many of the policies followed by China under Mao's leadership reflect his particular interpretation of Chinese history, the protracted revolutionary struggle and his experiences in the liberated areas, as for instance, in the dialectical attempt to transform the physical environment by waging war against Nature.

The Party and the PLA won the peasant revolution virtually without Soviet help, but the ideological origins of communism could not be ignored. In 1950 the Sino-Soviet Treaty of Friendship, Alliance and Mutual Assistance was signed, and then the First Five Year Plan, 1953–7, implemented. Thus, a Stalinist, centralized, large-scale industrial orthodoxy was superimposed upon the indigenous, local-level agrarian philosophy. The tensions between the two strands have never really been resolved, even when China was supposedly successfully walking on two legs.

In 1958 the Great Leap Forward was launched by forcing the pace of industrialization through the regimentation of labour, and imposing agricultural communization. In fact, as a direct consequence both industrial and agricultural output dipped disastrously, and famine stalked the land. Mao's power was eclipsed temporarily. In 1960 the Soviets withdrew technical assistance, due to ideological differences, which resulted in subsequent public rift. Then, in 1966, Mao reasserted his authority in the abortive Cultural Revolution, mobilizing the Red Guards, who challenged many people in leading positions, persecuting especially the educated classes. In 1976 Mao died, to be succeeded by a more moderate leadership. Beyond 1978, an open door policy was instituted, accompanied by major economic reforms in the countryside, focusing on the household responsibility system, which effectively returned the land to private peasant hands.

Unlike the situation in Eastern Europe and the Soviet Union, economic reform was not matched by political reform, and any hope of immediate liberalization was extinguished for the time being in the events of Tiananmen Square, in June 1989. This has left China at a cross-roads, wholly undecided whether to allow further cautious economic reform, or to return to a centrally planned economy; in the event, there may well be a mix of the two models.

The ebb and flow of these momentous political movements has impinged upon every aspect of China's changing geography, and, in turn, in many and diverse ways, China's geography has affected the course of events in contemporary political history, as will be evident below.

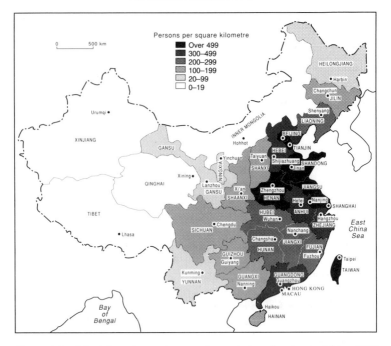

*Figure 5.1* Administrative regions and population density in China 1990

## POPULATION: THE PREDOMINANT FACTOR IN NATIONAL DEVELOPMENT

The 1953 Chinese census registered a total of 583 million. In the next thirty years the population almost doubled, for by the 1982 census the tally topped 1 billion. Between 1982 and 1990, when the fourth census was conducted, the population expanded by over 125 million, an average of nearly 15.7 million or 1.48 per cent annually, to reach 1,133,682,501, with a density of 118 per square kilometre. Although family planning has slowed the rate of natural increase overall, by allegedly preventing 200 million births in the last twenty years, both the birth and natural growth rates actually rose slightly from 1982 to 1990, partly because the huge population base has contributed to a third baby boom, still underway. In 1990 the birth rate was 20.98 per thousand, the death rate 6.28 per thousand and the rate of natural increase 14.70 per thousand (see Table 5.1).

The sex ratio was 106.6 males per 100 females (106.3:100 in 1982). Han Chinese accounted for 91.96 per cent of the total, and national

*Table 5.1* Population distribution, densities, vital statistics and urbanization (by major administrative unit) in China 1990

| Administrative unit | Total | Density (sq km) | Birth rate (0/00) | Death rate (0/00) | Natural increase | Urban % |
|---|---|---|---|---|---|---|
| *Provincial-level municipalities* | | | | | | |
| Beijing | 10,819,407 | 644 | 13.35 | 5.43 | 7.92 | 73.08 |
| Shanghai | 13,341,896 | 2118 | 11.32 | 6.36 | 4.96 | 66.23 |
| Tianjin | 8,785,402 | 777 | 15.50 | 5.98 | 9.52 | 68.65 |
| | | | | | | |
| *Provinces* | | | | | | |
| Anhui | 56,180,813 | 404 | 25.04 | 5.79 | 19.25 | 17.90 |
| Fujian | 30,048,224 | 248 | 23.45 | 5.70 | 17.75 | 21.36 |
| Gansu | 22,371,141 | 49 | 22.85 | 5.92 | 16.93 | 22.04 |
| Guangdong | 62,829,236 | 353 | 21.96 | 5.34 | 16.62 | 36.77 |
| Guizhou | 32,391,066 | 184 | 23.77 | 7.13 | 16.64 | 18.93 |
| Hainan | 6,557,482 | 193 | 22.95 | 5.22 | 17.73 | 24.05 |
| Hebei | 61,082,439 | 325 | 19.66 | 5.76 | 13.90 | 19.08 |
| Heilongjiang | 35,214,873 | 78 | 17.51 | 5.33 | 12.18 | 47.17 |
| Henan | 85,509,535 | 512 | 24.03 | 6.18 | 17.85 | 15.52 |
| Hubei | 53,969,210 | 290 | 24.32 | 6.84 | 17.48 | 28.91 |
| Hunan | 60,659,754 | 286 | 24.03 | 7.07 | 16.96 | 18.23 |
| Jiangsu | 67,056,519 | 654 | 20.54 | 6.07 | 14.47 | 21.24 |
| Jiangxi | 37,710,281 | 226 | 24.47 | 6.59 | 17.88 | 20.40 |
| Jilin | 24,658,721 | 132 | 18.40 | 6.12 | 12.28 | 42.65 |
| Liaoning | 39,459,697 | 270 | 15.60 | 6.01 | 9.59 | 50.86 |
| Qinghai | 4,456,946 | 6 | 22.65 | 6.84 | 15.81 | 27.35 |
| Shaanxi | 32,882,403 | 160 | 23.49 | 6.49 | 17.00 | 21.49 |
| Shandong | 84,392,827 | 539 | 18.86 | 6.25 | 12.61 | 27.34 |
| Shanxi | 28,759,014 | 184 | 22.31 | 6.25 | 16.06 | 28.72 |
| Sichuan | 107,218,173 | 188 | 17.78 | 7.06 | 10.72 | 20.25 |
| Yunnan | 36,972,610 | 94 | 23.59 | 7.71 | 15.88 | 14.72 |
| Zhejiang | 41,445,930 | 407 | 14.84 | 6.10 | 8.74 | 32.81 |
| | | | | | | |
| *Autonomous regions* | | | | | | |
| Guangxi | 42,245,765 | 178 | 20.71 | 5.96 | 14.75 | 15.10 |
| Inner Mongolia | 21,456,798 | 18 | 20.12 | 5.79 | 14.33 | 36.12 |
| Ningxia | 4,655,451 | 90 | 24.56 | 5.07 | 19.49 | 25.72 |
| Tibet | 2,196,010 | 1.8 | 27.60 | 9.20 | 18.40 | 12.59 |
| Xinjiang | 15,155,778 | 9 | 24.67 | 6.39 | 18.28 | 31.91 |

*Source:* 1990 Census.

minorities for the remaining 8.04 per cent. Households numbered nearly 277 million, averaging 3.96 persons (4.41 in 1982). Between 1982 and 1990, the urban population grew from 206.6 million to 296.5 million, or from 20.6 to 26.23 per cent. Significantly, there was a floating population of nearly 30 million – almost certainly an undercount, as the real figure for the number of transients could be at least double this total, possibly in the order of 80 million.

Population composition dominates every aspect of China's development, but its impact is hotly debated. Maoist precepts, which hold that a large population remains a source of national pride and strength, continue to appear.

> Powerful political and economic pressures from foreign countries have been unable to overwhelm China since 1949. This is mainly because of socialist China's large population. The advantage of a large country is that it can concentrate its human and financial resources on running businesses that are beyond the capabilities of small, economically backward countries. ... For a large, populous country, however, there are prerequisites for the demonstration of its political and economic advantages. First, there must be a cohesive force. Otherwise, the larger the population, the more it is 'like a sheet of loose sand,' in a state of disunity. The most important factor for a cohesive force is the leadership of the Communist Party. Secondly, China ... needs to seek common prosperity for its 1.1 billion people. This can only be gained by taking the socialist road. Thirdly, it is necessary to have a talented population.
>
> (*Qunyan* [*Opinions of the masses*], no. 1, 1990)

One counter-argument concerns resources: 'About 20 per cent of all annual increases in the national income has to go toward fulfilling just the basic needs of the extra population.' Moreover, 'China has vast supplies of natural resources, but these too are being consumed at an alarming rate. ... By world standards, China's available water supplies can realistically support only 250 million people, and as for arable land, only 330 million people can be adequately fed' (*China Today*, March 1990; hereafter, *CT*). Such figures are arbitrary and probably exaggerated, yet problems associated with over-exploitation of resources, compounded by environmental degradation and pollution are widespread.

In any event, China is determined to hold population at 1,250 million in the year 2000, whilst conceding that this will be very difficult. Therefore, effective birth control is seen to be a paramount necessity if China is to meet the strategic target of socialist modernization.

### Family planning: the single- or two-child family?

Despite the transparent urgency to devise and implement – even to impose – an effective family planning strategy, over the decades under review there have been repeated shifts in policy. Initially, the Maoist position was adopted, which argued that there was strength in numbers. Tere was no mention of birth control in the First Five Year Plan (1953–7), yet the opening campaign was launched in 1956. The explanation for this apparent contradiction was that it was in the best interests of the mother and child that births should be neither too early nor too frequent nor too many. The initial campaign mainly targeted the urban areas and was too short-lived to have had any real effect, collapsing under the impact of the Great Leap Forward in 1958.

A second family planning campaign began in the early 1960s, in the aftermath of the serious socio-economic chaos engendered by the Great Leap. Intensified propaganda for delayed marriages and an improved distributional network for new methods of birth control (such as the IUD) were nullified by the advent of the Cultural Revolution in 1966. Thus, neither of the earlier campaigns had time to have much direct effect on population growth (Freeberne 1964).

A third, sustained, yet fluctuating family planning campaign has now been in place for the last two decades. To start with, it was essentially an extension of the earlier programmes, but with the addition of the contraceptive pill. Then, from the early 1980s, the single-child policy was promoted, employing ever tougher measures. At first, stress was placed on the incentives for those families pledging to have just one child. These varied over time and from place to place, but included modest financial rewards, offers of improved accommodation and better educational and medical provision – even permission to queue-jump. Generally, the penalties exacted consisted of a loss of these privileges. Over-zealous harridans on the street committees, the so-called 'grannie police', kept a constant watch on their charges; forced abortions for those breaking set quotas or conceiving out-of-turn became commonplace. Similarly with infanticide in rural areas, where girl children in

*Table 5.2* Main national minorities with populations of over 500,000 in China 1982–90

| Nationalities | 1982 ('000) | 1990 ('000) | Growth 1982–90 (%) |
|---|---|---|---|
| 1 Zhuang | 13,388 | 15,490 | 15.7 |
| 2 Manchu | 4,304 | 9,821 | 128.2 |
| 3 Hui | 7,227 | 8,603 | 19.0 |
| 4 Miao | 5,036 | 7,398 | 46.9 |
| 5 Uygur | 5,963 | 7,214 | 21.0 |
| 6 Yi | 5,457 | 6,572 | 20.4 |
| 7 Tujia | 2,835 | 5,704 | 101.2 |
| 8 Mongolian | 3,417 | 4,807 | 40.7 |
| 9 Tibetan | 3,874 | 4,593 | 18.6 |
| 10 Bouyei | 2,122 | 2,545 | 19.9 |
| 11 Dong | 1,426 | 2,514 | 76.3 |
| 12 Yao | 1,404 | 2,134 | 52.0 |
| 13 Korean | 1,766 | 1,921 | 8.7 |
| 14 Bai | 1,132 | 1,595 | 40.9 |
| 15 Hani | 1,059 | 1,254 | 18.4 |
| 16 Kazak | 908 | 1,112 | 22.4 |
| 17 Li | 818 | 1,111 | 35.8 |
| 18 Dai | 841 | 1,025 | 22.0 |
| 19 She | 369 | 630 | 70.9 |
| 20 Lisu | 481 | 575 | 19.5 |
| 21 Gelo | 54 | 438 | 714.1 |
| 22 Han | 940,880 | 1,042,482 | 10.8 |
| Total | 1,008,175 | 1,133,683 | 12.5 |

*Note:* An additional thirty-five nationalities were identified in the 1990 census; of these, thirteen had populations of between 100,000 and 0.5 million each, and twenty-two populations of below 100,000 each; other unidentified minorities totalled 881,838.

particular were drowned. There was evidence of seriously skewed sex ratios in many country areas, and worries were expressed regarding the 'little emperors', or badly spoilt youngsters growing up in the towns. The widespread emergence of such phenomena was behind the relatively recent decision to relax family planning policies by allowing a second child to rural families.

New family planning legislation covers the reasons for implementation; methods of birth control; the upbringing of healthy children; administrative procedures for extending the national programme and specific measures to award and punish citizens. Until recently, the national minorities have been exempt from family planning, but now they too are more often included in local plans. Undoubtedly, this reflects Chinese concern over the rapid expansion of minority numbers and their growing sense of nationalism, as from 1982 to 1990, the Han Chinese increased by 10.80 per cent, whilst the national minorities grew at 35.52 per cent (see Table 5.2).

## National minorities: a great family of unity and fraternity

There are over 91.2 million people living in China who are not themselves Chinese, but who belong to one of fifty-five national minorities. This fact alone suggests a potential for friction, especially when it is considered that the national minorities are mainly dispersed throughout the peripheral areas, removed from central authority in Beijing, relatively remote and economically backward. Since 1949, these tensions – based on marked racial, linguistic and religious differences – have surfaced repeatedly, in the form of conflict between 'great Han chauvinism' and 'local minority nationalism', notably in Tibet and Xinjiang.

The People's Liberation Army, for example, marched into Tibet in 1950; to some the soldiers came as liberators, to others as invaders. In 1959 the Tibetans rose in revolt and their spiritual leader, the Dalai Lama, fled into exile, since when the volatile situation in Tibet has continued to simmer. The supposedly advantageous status of an autonomous region was not granted until 1965. In the mean time, large numbers of Chinese settlers flooded in, often occupying positions of authority and enjoying comparative affluence and prestige, leaving the Tibetans depressed, second-class citizens. Serious anti-Chinese disturbances in 1988, led to skirmishing and deaths and the imposition of martial law in Lhasa between March 1989 and May 1990.

Xinjiang is a vast territory occupying one-sixth of China, but its multi-ethnic population totals only 15 million, although augmented by many Chinese immigrants from the 1950s onwards. In 1962 national minority factions rebelled against the Chinese, calling for the establishment of an independent East Turkestan Muslim Republic – in direct defiance of the Chinese constitution, which confirms China as a unitary state. Nomads in their tens of thousands crossed the long, mountainous border with the Soviet Union, hoping to join their cousins on the other side. The Soviet Union then attempted to seal the still disputed border. Feuding along the north-western frontier has been ongoing, with countless border violations on both sides (Freeberne 1966; 1968). However, unlike the north-eastern frontier in Heilongjiang, where war broke out in 1969, war has not been declared. In 1988 Sino-Soviet negotiations to resolve both frontiers were resumed. But inter-ethnic relations have remained extremely tense in Xinjiang, with further major unrest in mid-1990.

## IDEOLOGICAL AND PRAGMATIC ECONOMIC CHANGE

Economic planning, which has brought about numerous changes in the economic geography of China – for all its initial asperity and apparently unyielding ideological dogmatism – has itself undergone many shifts.

The years from 1949 to 1952 were a time of reconstruction, when the war-torn nation was set painfully, and a little unsteadily on to its feet, and most major pre-Liberation production indices were bettered. Under the First Five Year Plan, 1953–7, the Chinese aimed 'to convert China, step by step, from a backward, agricultural country into an advanced, socialist, industrial state' (*First five-year plan for development of the national economy of the People's Republic of China in 1953–1957*, 1956). However, in the process of giving undue support to the industrial sector, which received 61.8 per cent of investment, agriculture was very seriously under-resourced, with a mere 6.2 per cent of funding. Through this gross miscalculation, the seeds of the subsequent national famine disasters were sown, and the masses were soon to find themselves 'eating bitterness', rather than grain.

Heavy industry remained at the heart of the Second Five Year Plan, which ran in name only between 1958 and 1962, and the continuation of the essentially Stalinist, guns-before-butter model was also to provide the launch-pad for the Great Leap Forward in 1958 – a cataclysmic and fateful year which saw the communization of agriculture as well. The Great Leap was a time of phrenetic activity, sorely overstretching the physical strength of the population, which was to slump exhausted through working too hard and from hunger. The cumulative effect was glaringly obvious in terms of production statistics. Although starting from a low level, between 1949 and 1952, industrial output had grown on average by 27 per cent annually. By 1960 it was only 4 per cent, and much of the claimed industrial output was virtually valueless.

For example, the widely acclaimed backyard iron and steel furnaces produced metal which was only of use in tipping agricultural implements.

The winter of 1960–1 was the nadir in China's economic fortunes. Industrial production more or less ground to a halt, and millions starved, as a result of harvest shortfalls, due in part to bad weather, but greatly magnified by the excesses of communization. Famine, supposedly banished once and for all after Liberation, had returned. From 1963 to 1965 China's economy depended on annual accounts, effectively functioning outside the framework of a five year plan, a clear indication that there were serious economic problems. The rigours of the 1960–1 winter dictated a reversal in planning emphasis, with an all-out effort to save agriculture. By raising work incentives and easing work pressures on the peasants, and by channelling industrial production into making items like fertilizers, tractors and irrigation and drainage equipment, there were the first signs of striking a more rational balance in the economy as a whole.

Then, as the economy was beginning to recover, and just as the Third Five Year Plan, 1966–70, was getting underway, along came the Great Proletarian Cultural Revolution, 1966–76, a movement which was to result in even deeper traumas than the Great Leap, as the more reasonable economic policies were swept aside: now, better Red than Expert. In fact, politics and economics became hopelessly enmeshed. During the opening rounds of the Cultural Revolution, many millions deserted their production posts and, waving their little red books, formed a sycophantic ocean around the Great Helmsman, Mao Zedong, and a vermilion sea of politically orchestrated protest, aimed at demons both domestic and foreign, no doubt ideologically inspiring, but very damaging to economic output.

In the name of self-reliance and self-sufficiency, important economic features during the Cultural Revolution included, for instance, an over-emphasis on the production of staple crops, concentrating on stable, high-yield fields, at the expense of the diversification of agriculture. Also, the extraordinary promotion of an agricultural model, the Dazhai (Tachai) production brigade, Shanxi, which the entire countryside was to emulate after 1964, together with the complementary industrial model of Daqing (Taching), the oil field in Heilongjiang, which was to inspire the urban workforce to ever greater efforts. By some curious alchemy, each grain harvested was transmuted into a bullet to be fired at the enemy, which might just as well be a Soviet as an American soldier. Subsumed under, and swamped by politics, economics hardly stood a chance, and China is still calculating the cost of the Cultural Revolution on innumerable fronts.

## The command economy: blind to spatial diversity

In view of the vast and varied territory to be tamed and brought into maximum production, it is quite remarkable that throughout the last forty years, the socialist command economy, based in the northern capital, Beijing, should have attempted in blinkered fashion to impose more or less common economic planning directives on the entire country, without appropriate recognition of, and regard for, the special geographical characteristics of both adjacent and distant regions. This surely defied logic, as there never could have been a single, comprehensive set of formulae which would work equally well for all sectors of the economy and in all parts of a country of such immense and rich geographical diversity. For instance, how could it have been that agricultural collectivization would work equally well in the padi of the fertile Zhujiang (Pearl) River Delta, the environmentally threadbare, semi-arid and arid north-west or the nomadic, pastoral lands of Inner Mongolia?

Simply put maybe: but as a result of politically inspired, short-sighted, unimaginative and rigid central economic planning, there has been an ever-present, debilitating and damaging tension between the centre, Beijing (to include other growth poles like Tianjin and Shanghai) – attracting disproportionate investment – and the neglected middle-distance regions and the periphery.

Before 1949, industry was concentrated mainly in Shanghai, Manchuria and Tianjin. A number of factors dictated the dispersal of subsequent industrialization. First, the existing coastal and north-eastern locations were an unwelcome reminder of the imperialist heritage. Second, the over-concentration of industrial plant was dangerous from a strategic point of view. Third, a more rational distribution could be achieved by siting new industries near to raw materials, energy sources, labour and markets. Finally, there was a desire to develop the economy of backward areas.

Despite the determination of the Chinese to move the heart of Chinese industry inland and away from the coast, however, the eastern seaboard and the north-east have persisted in exerting something of a strangle-hold on the location of industry for the last forty years. Indeed, the pull of the coast has strengthened rather than diminished under the post-1978 reforms – for on to what should the open-door of reform open but on to the outside world? Furthermore, access surely had to be through the front rather than the back door, through Shanghai rather than some inland, frontier city? Be this as it may, pre-1978 efforts to break the coastal dominance resulted in no more than a piecemeal and extremely patchy effect – islands of industry set in a rural sea.

To illustrate with both early and late examples. Three specific locational requirements were outlined in the First Five Year Plan, 1953–7. The most important task was to complete the major part of the north-eastern industrial base centred on the integrated iron and steel works at Anshan. In association with this reconstruction, work was to be undertaken at the coal-mining centres of Fushun, Fuxin and Hegang, the iron and steel works at Benxi, the machine-building works at Shenyang, and the power installations in Jilin. But other regions also began to emerge. For example, in north China, apart from the growth of Baotou, Tianjin and Beijing, the Taiyuan and Shijingshan iron and steel centres were developed. In the north-west, Xian and Lanzhou grew in importance, and there was expansion of Gansu's Yumen and Xinjiang's Karamay oil fields. Meanwhile, light and handicraft industries were encouraged on an even wider basis. The rise of central and especially southern centres was much slower, although cities as far apart as Hangzhou, Daye, Liuzhou, Chengdu, Chongqing, Kunming, and Guangzhou started to industrialize and were earmarked for future expansion.

In this early phase in the long march towards modernization, forces of industrial inertia were already running counter to the plans for the dispersal of Chinese industry, and before the Cultural Revolution began in 1966, far greater attention had been given to the established industrial regions in the north-east, north and east of the country than to the newly emerging regions. Of no single location has this been more true than of Shanghai, which for over forty years has prospered as by far the largest, potentially most strategically vulnerable, and ideologically most irritating nerve centre in the Chinese industrial anatomy. Shanghai typifies the worst sort of run-away urban growth, and its development has been contrary to at least some Chinese thinking on industrial location, such as the attempt at the integration of town and countryside in preference to the swallowing up of rural areas by unremitting urbanization (Freeberne and Fung 1981).

The official, catch-all locational policy at the time of the Cultural Revolution – optimistically expressed and including a nice illustration of the contemporary siege, anti-urban mentality, together with the sidewinder aimed at the Soviet Union – was that

> the geographical distribution of China's industry has been brought into accord with the policy that a nationwide dispersal of enterprises should be accompanied by regional concentration. ... China's industrial construction has also pursued the policy of integration of industry and agriculture and of town and countryside. Moreover, *to prevent the emergence of revisionism and peaceful evolution to capitalism* we have adhered to the policy of not building new cities, big cities and high-standard living accommodation. Many newly built factories are scattered in the rural areas.
>
> (*Peking Review* 29 September 1967) [emphasis added]

For, in the aftermath of the Great Leap Forward, planning priorities had shifted to allow, first, for an all-out effort to save agriculture, and, somewhat later, parallel development of the two main sectors.

Beyond the Cultural Revolution, under the post-1978 economic reforms, there evolved an unabashed emphasis on the further advance of the coastal belt, largely at the expense of the remainder of the country. In particular, four Special Economic Zones (SEZs) were set up in the south-east at Shenzhen, Zhuhai, Shantou and Xiamen in 1980, and a fifth in 1988, on Hainan. Also, fourteen coastal ports were designated open cities in 1984. The guiding philosophy behind the creation of the SEZs has been to combine the advantages of imported foreign capital and technology with cheap but often unskilled, indigenous labour and plant, fortified by a topping of generous tax concessions.

At the start of 1990, authorities from the seven coastal provinces met in Hangzhou, to discuss the fate of over 12 million township small industrial enterprises – feeling the squeeze from post-1988 central government retrenchment – as in recent years, a majority of such units have concentrated in these provinces, accounting for nearly 60 per cent in gross production value of all township enterprises nationwide. The delegates urged Beijing to resuscitate the stable and sustained development of all township enterprises, which was directly related to the standard of living enjoyed by millions of workers in the countryside. Strikingly, the meeting warned that otherwise there was a very real danger of unrest and a threat to political stability. However, no new policies emerged from the session, which simply echoed current Beijing thinking that there was a need for township enterprises generally to move away from supplying the domestic market to export-orientated output, to complement state-run industries, as well as agriculture – another catch-all formula, easier to dream-up than to implement.

The publicity given to the inauguration of the Pudong Industrial Zone in Shanghai in 1990 intensified the competition for scarce foreign capital, even harder to come by after Tiananmen. One cartel sought to form an 'economic base', linking Guangzhou, Shenzhen, Huizhou and Hong Kong; should it manage to take off, it could be very influential during the 1990s. Huizhou, for example, 'will be a key player in the export-orientated economy of south China' once the base is fully established, specializing in electronics, chemical and machine industries. Major projects at present under construction within the confines of the base include an oil refinery, a large electronics plant and a car plant in Huizhou; an airport and deepwater berths in Shenzhen; the controversial Daya Bay nuclear power plant and the super-highway linking Hong Kong, Shenzhen and Guangzhou (*China News Service* 7 May 1990).

Under the Eighth Five Year Plan, 1991–5, China is looking to unlock the rich natural treasure-trove of the remote, immense and economically backward western half of the country, without putting an additional brake on the expansion of the coastal zone. Over the years, widespread discontent has surfaced in response to what the several major outlying regions nationwide regard as neglect by central government. The scheme to exploit the natural wealth of the west is intended to counter allegations of favouritism and largesse by Beijing. Steps to open up the west include a sizeable increase in money to employ labour from poverty-wracked areas in highway and water conservation projects; the planned transfer of funds generated as the state surrenders direct control over much of the military industry, previously located in the west for strategic reasons, allowing it to merge with the local economy; and the recent opening of cross-border trade with neighbouring countries, including Russia and Central Asia.

China's gross industrial output in 1990 totalled 2,385.1 billion *yuan*, an increase of 7.6 per cent over 1989. Selected items are included in Table 5.3.

*Table 5.3*  Industrial and energy output in China 1990 (selected items)

| Item | Output | Increase/decrease on 1989 (%) |
|---|---|---|
| Steel | 66.04 mn tons | +7.2 |
| Rolled steel | 51.21 mn tons | +5.4 |
| Chemical fertilizers | 19.12 mn tons | +6.1 |
| Machine tools | 0.1178 mn | −34.1 |
| Motor vehicles | 0.5091 mn | −12.8 |
| Tractors | 0.039 mn | −1.5 |
| Cotton yarn | 4.5 mn tons | −5.6 |
| Cotton cloth | 18,000 mn metres | −4.9 |
| Woollen fabrics | 280 mn metres | stet. |
| Household refrigerators | 4.754 mn | −29.2 |
| Household washing machines | 6.526 mn | −20.9 |
| Television sets | 26.62 mn | −3.8 |
| Of which colour sets | 10.23 mn | +8.8 |
| Tape recorders | 29.7 mn | +26.4 |
| Cameras | 1.899 mn | −22.6 |
| Bicycles | 31.41 mn | −14.6 |
| Cigarettes | 32.9 mn cases | +3.0 |
| Cement | 203 mn tons | −3.3 |
| Timber | 54 mn cu m | −6.9 |
| Power generating equipment | 11.43 mn kilowatt | −2.7 |
| Coal | 1,080 mn tons | +2.5 |
| Crude oil | 138 mn tons | stet. |
| Electricity | 618,000 mn kwh | +5.7 |
| Of which HEP | 126,000 mn kwh | +6.5 |

*Note:* Pronounced fluctuations in performance, with tape recorders up by 26.4 per cent and machine tools down by 34.1 per cent are partly explained by the post-1988, anti-inflation, economic austerity programme.

## Energy

During the Opium Wars, China sought to defeat the enemy by prohibiting the export of rhubarb, ensuring that the foe died in knotted agony. The overriding problem with the economy is that it suffers from both ideological and physical constipation, the one often causing the other. Hence, the ever-present bottle-necks are best illustrated by an inadequate transport and communications network, and by a perpetually stuttering energy supply.

As the world's third largest energy producer, China is an energy giant, yet badly crippled. Whilst occupying second place as a coal producer, fifth place in oil and sixth place in installed hydro-electric power (HEP) capacity in world tables, the central problem is not so much one of potential energy availability as one of actual distribution, given China's size and the long transport hauls involved. For example, China's immensely rich coal reserves are mainly found north of the Changjiang, whereas the southern half of the country lacks good quality supplies. An enormous infrastructure is required to mine, process and transport coal. For instance, 7 million to 8 million coal trucks are needed annually to carry the coal to the consumer, placing tremendous strains on the rail network. Oil production is even more localized and mainly situated distant from the leading economic areas, whilst the huge, largely untapped HEP potential is again in remote, economically backward parts of the country. Similarly, China's first allegedly wholly indigenous nuclear power station at Quinshan, 126 kilometres from Shanghai, which went into production in 1991, will serve only a limited hinterland.

As well as opening up new coal fields, the increasing mechanization of existing mines is planned. At present, 58 per cent of the work carried out in state-run mines is mechanized. During both the Great Leap and the Cultural Revolution, when a great deal of effort went into searching for coal south of the Changjiang, local sources were tapped, to support the growing network of small-scale rural industries (Freeberne 1971). 'Since the early 1980s, many small local mines have also gone into business, and at present about half of the country's coal output is from small [mines] run by townships, villages and even individuals' (*CT* February 1990).

Reminiscent of the policy of walking on two legs – combining large-scale, capital intensive, advanced technology and small-scale, labour intensive, low technology units – such parallel developments underline both strengths and weaknesses in the Chinese development model. For example, on the one hand, small rural units help to address the question of chronic underemployment in the countryside; on the other hand, they add to the problem of pollution. In addition, of the 1 billion tons of coal consumed in 1989, about 20 per cent was wasted, due to inefficient use (*Xinhua News Agency* 18 February 1990; hereafter *XNA*).

## Agriculture: the foundation of the national economy

Startlingly, not much more than 10 per cent of the land is in agriculture, so that China must feed almost a quarter of the world's peoples on only 7 per cent of the world's arable area. Initially, the government claimed that the cultivated area could be more than doubled, but since 1949 the arable area has decreased rather than increased, as land lost to agriculture through competition from housing and industry has expanded at a greater rate than the amount reclaimed. Deserts, too, account for roughly 10 per cent of the land surface, and desertification is encroaching on agriculture.

Agriculture occupies the predominant role in China precisely because it not merely has to feed the population, but also must produce commercial and industrial crops and a sufficient surplus for industrial investment, to the extent that the character, pace and degree of economic progress will continue to be strongly influenced by agricultural performance, to which there are finite limits.

## Collectivization: mutual aid team to rural people's commune

No aspect of the agricultural sector is of more significance than the system of organization. Traditionally, peasant life in the old China before 1949 was characterized as brief, bitter and brutal. Most importantly, the vast majority of peasants were either tenant farmers or landless labourers, mainly abysmally poor and working tiny parcels of land. Prior to Liberation, the landlord and rich peasant classes, totalling less than 10 per cent of the rural population, owned over 70 per cent of the cultivated land. Under the land reform movement this land was seized and redistributed to the poor and middle peasants, an essential factor in the winning of widespread rural support by the communists. By 1952 300 million landless peasants had been given 700 million *mou* (a *mou* equals one-fifteenth of a hectare) of land, as well as homes, draught animals and farm implements, and accumulated revenues owed to the landlords were waived, although new government taxes were introduced. Land reform was completed with relatively little bloodletting.

Mutual aid teams (MATs) were established. By 1952, 40 per cent of the peasant households belonged to such teams and by 1954, 58 per cent were members of about 10 million teams. The main feature of the MATs was the pooling of labour, tools and animals, mostly at seed-time and harvest, in the case of temporary teams, but on a more general basis where permanent MATs were founded. Most property remained in private hands, and the teams usually consisted of up to about ten households. Parallel with the later stages of the spread of the MATs, permanent MATs were consolidated into

Agricultural Producers' Co-operatives (APCs). In 1952 there were 3,600 APCs, and by the middle of 1955, there were 670,000, covering 17 million peasant households. By the end of 1956, 120 million peasant households, representing 96 per cent of the rural population had joined APCs. Under the lower stage APCs, the land was owned privately, but pooled and run by a central management. Peasant income depended upon labour and land contributions, and each APC included about thirty or forty households.

Transition from lower stage to higher stage or advanced APCs occurred chiefly in 1956, when by sleight of administrative hand land became the property of the APC, together with draught animals and large tools, but small private plots were retained by individuals. Labour was divided into production brigades, and sometimes production teams, brigades averaging about twenty households. The average number of households in a higher APC was 160, and the average land holding 378 acres. The comprehensive sweep of the operation, the crushing of all peasant opposition, and the speed at which collectivization was implemented were quite staggering. Mao Zedong gave the movement tremendous impetus in July 1955 with his famous address *On the question of agricultural co-operation*, in which he chastised wavering cadres: 'Throughout our countryside a new upsurge in the socialist mass movement is in sight. But some of our comrades are like a woman with bound feet, tottering along and constantly grumbling to others: "You're going too fast."'

Then came the final step, lifting collectivization on to a new and higher plane, when 'the three red banners' of the general line, the Great Leap Forward and the establishment of the communes were unfurled. In this way Mao Zedong's recommendation that the new unit of organization should combine agriculture, industry, trade, education and the militia was adopted. By the end of September 1958, 26,425 communes, embracing 122 million or 98.2 per cent of the peasant households, had been set up. By September 1959, consolidation reduced the number of communes to 24,000, averaging 5,000 households each. The communes contained in all 500,000 production brigades, usually equivalent to the old advanced APC and averaging between 200 and 300 families. In addition, there were 3 million production teams, with up to forty households each. In 1964 the number of communes was increased to 74,000; but by the time that the commune system was dismantled in the early to mid-1980s there were about 54,000, the fluctuating numbers indicative of the fact that at no stage was there unanimity as to the optimum size of the commune.

Apart from planning the agricultural year, the commune was responsible for all other economic activity, as well as social functions such as assuring the seven guarantees to do with food, clothing, medical care and childbirth, education, housing, marriages and funerals. Initially, there were undoubted excesses, under which private property was virtually abolished; corvée-type work projects, political study and para-military training often were too demanding; and family life was disrupted by replacing home-cooking with inadequate communal mess halls and introducing a crèche system, which might remove children from their parents for unreasonable periods. After the opening trial months, conditions were relaxed, but this was too late to prevent the frantic pressures of communal life from being reflected in the statistical chaos of the Great Leap.

Undoubtedly, some communes situated in kinder climes prospered, but in all probability this was despite rather than because of the system. The supposed advantages of scale did not always materialize, as evidenced by the repeated redrawing of commune boundaries; and the replacement of the communes by the responsibility system represented a whittling away of and the ultimate reduction in the optimum size of production unit. But above all else, the Chinese failed to devise an effective method of motivating the peasants by rewarding effort and providing appropriate incentives. For over twenty years Chinese planners indulged in expensive tinkering with the communes – as much as anything, largely in response to dogma, and so as to save ideological face. A more satisfactory package of remuneration came only with the introduction of the responsibility system.

Agricultural output totalled 185 million tons of grain in 1957, the last year of the First Five Year Plan. Then, for several years, there was considerable uncertainty regarding Chinese harvests, as there was a combination of statistical blackout and a collapse of the national accounting system. Indeed, the first figures for 1958 did represent a great leap as an official claim of 375 million tons of grain was registered, although later revised by prime minister Zhou Enlai to 250 million tons. Even the revised total, as well as the 270 million tons for 1959, were believed to have been heavily inflated. Tentatively, actual production may have been more at the level of about 200 million tons in 1958, 170 million tons in 1959, and 143 million tons in 1960 (see Table 5.4).

## Household responsibility system: whose responsibility?

The new household responsibility system, which has resulted in a marked growth in agriculture, has also been tested in the industrial sector, but has been much less successful as an industrial tonic. Agriculture was decommunized by the early to mid-1980s, and replaced, to begin with, by a short-term contracting system in which small groups, family units or individuals leased land and agreed to meet production quotas; they were then encouraged to off-load any surplus produce, generally in the free markets, once again officially promoted. Subsequently, contracts were issued for much longer periods of up to thirty years, even permitting niceties such as inheritance rights.

As both government and local confidence in the responsibility system grew and agricultural output expanded, there was a substantial increase in specialist production, such as industrial crops, vegetables, fruit or livestock (see Table 5.5). One striking regional example is found in the Zhujiang River Delta where fish ponds abound. The pond banks are planted with mulberry, sugar cane

*Table 5.4*   Grain output in China (selected years)

| Year and comment | Production (million tons) |
|---|---|
| Pre-1949 peak | 139 |
| 1957 Last year of the First Five Year Plan | 185 |
| 1958 Great Leap Forward, communization | 200 |
| 1959 Second year of Second Five Year Plan | 170 |
| Middle year of 1960 Second Five Year Plan | 143 |
| 1966 Cultural Revolution Third Five Year Plan | 214 |
| 1970 Last year of Third Five Year Plan | 239 |
| 1975 Last year of Fourth Five Year Plan | 284 |
| 1978 Beginning of economic reforms | 304 |
| 1984 Penultimate year of Sixth Five Year Plan | 407 |
| 1985 Last year of Sixth Five Year Plan | 379 |
| 1989 Penultimate year of Seventh Five Year Plan | 407 |
| 1990 Last year of Seventh Five Year Plan. A record harvest, but the average per capita consumption of 370 kg was 30 kg lower than in 1984 | 425 |
| 2000 Target | 500 |

*Note:* Grain production is one issue; average per capita grain consumption is another matter entirely. Overall since 1949, there have been marked improvements in distribution, apart from the famine in the early 1960s. There was discontent amongst the peasants in 1989 and 1990, however, when the government was forced to issue IOUs to pay for state grain quotas.

and bananas; the mulberry is fed to silkworms, and pupae and droppings to the fish; pond mud is then applied as fertilizer along the dykes. This move away from an over-emphasis on staples towards diversification has resulted in ever more complex land-use patterns, a trend which the authorities must regulate carefully lest the swing from grains becomes too great, and food shortages threaten.

*Table 5.5*  Agricultural output: major cash crops, animal products and livestock in China 1990

| Product | Million tons | Increase over 1989 (%) |
|---|---|---|
| Cash crops | | |
| Cotton | 4.47 | 18.1 |
| Oil-bearing crops | 16.15 | 24.7 |
| Of which: | | |
| Rapeseed | 6.93 | 27.5 |
| Sugarcane | 57.27 | 17.4 |
| Sugarbeet | 14.53 | 57.2 |
| Jute, ambary hemp | 0.72 | 9.5 |
| Cured tobacco | 2.26 | −6.2 |
| Silkworm cocoons | 0.53 | 9.4 |
| Tea | 0.53 | −0.3 |
| Fruit | 18.76 | 2.4 |
| | | |
| Animal products and livestock | | |
| Pork, beef, mutton | 25.04 | 7.7 |
| Milk (cow) | 4.13 | 8.2 |
| Wool (sheep) | 0.24 | 1.8 |
| Pigs slaughtered | 310,000,000 head | 6.2 |
| Pigs in stock | 360,000,000 head | 3.0 |
| Sheep and goats in stock | 210,000,000 head | −0.8 |
| Large animals in stock | 130,000,000 head | 2.7 |

*Source:* State Statistical Bureau (22 February 1991)

Major problems requiring urgent attention in the 1990s include, first, the persistent, chronic decline in arable land, due to competition from other forms of land use, such as industry and the ostentatious expansion of village housing, as often as not eating away at prime land. Second, the rapidly deteriorating agricultural infrastructure, as upkeep has been seriously neglected and commune-maintained projects have fallen into disrepair in the competitive quest for quick returns, under reform and the responsibility system. Third, the ideologically fraught question of mechanization: how can there be true modernization without extensive mechanization; but, as mechanization takes place, what will happen to the already burgeoning numbers of rural unemployed and underemployed?

Because of these and related questions, and in the light of current Chinese statements, some recollectivization cannot be ruled out, especially if there is a marked shift back to central planning in industry. Indeed, an amalgam of the responsibility system and partial recollectivization is quite probable.

Grain output is set to top 500 million tons by the year 2000. On the broad backs of the Four Modernizations (agriculture; industry; science and technology; defence), the Eighth Five Year Plan, 1991–5, and a new Ten Year Development Programme, 1991–2000, and by building socialism with Chinese characteristics, China seeks to quadruple gross national agricultural and industrial output (by value) between 1980 and 2000. By 1990, China appeared to be more or less on course, combined production having doubled in the previous decade. However, there was an ominous budget deficit in 1990 (officially 15 billion *yuan* [£1.6 billion], but using internationally accepted calculations, topping £10 billion), as China spent more in subsidizing prices and helping inefficient state enterprises than the total budget for education, science, culture, health and defence. Unless there was effective financial management political, economic and social stability might well be threatened, and the population trap sprung.

## ENVIRONMENTAL CHALLENGES: NO COUNTRY IS AN ISLAND

### Natural and/or man-made disasters

China enjoys a wide range of climates, hence maximizing the opportunity to grow a comprehensive list of crops. At the same time, given a monsoonal regime, China has been characterized as a land of alternatively too much or too little water, as well as traditionally a land of famine. Few, if any, years are clear of natural calamities.

For example, the period 1959 to 1961 is referred to as the 'three bitter years'. Indeed, there was evidence of serious flooding and droughts in 1959, and again in 1960. In 1960 it was claimed that half the land in agriculture was devastated. A retrospective assertion that the same was true for 1961 cannot be substantiated, however (Freeberne 1962). Later, three reasons were given for the failure of the Great Leap Forward: first, the withdrawal of Soviet technical assistance in 1960; second, bad weather and, lastly, 'mistakes' which the Chinese had made themselves, and which had turned natural disasters into a series of man-made calamities, costing over 20 million lives in widespread famines.

In July 1976 a major earthquake struck the coal-mining city of Tangshan, 7.8 on the Richter scale. More than 242,000 were killed outright and 165,000 'living dead' left behind, 'a horror story of mangled bodies, split skulls, collapsed buildings, looting, killing'. For years, the area was closed to foreigners, although the rehabilitation of Tangshan has been an apparent success (Qian Gang, *The Great China Earthquake*, 1990). In the old China natural disasters were thought to presage the death of an emperor who had lost the Mandate of Heaven, and in fact Mao Zedong died in September 1976. Also, after the Tangshan earthquake, rumours of impending earthquakes in various parts of China caused significant production losses, as workers stayed at home (*Guangming Ribao*, 19 October 1989). Over the last few years, the Chinese have developed methods of predicting earthquakes from observations of both animal behaviour and cloud formations.

Other more recent natural disasters made worse by human intervention include the major flooding in Sichuan in 1981, blamed on mismanagement of the upper streams of the Changjiang; and the forest fire in the Daxinganling, Heilongjiang in 1987, regarded by the Chinese as an ecological disaster of the first magnitude (Freeberne 1988). Lack of investment in water conservation and flood control, coupled with the failure to implement plans drawn up in 1981, were amongst the causes of the serious flooding in 1991, which again sparked off rumours of a change in leadership (Freeberne 1991).

Lately, Chinese meteorologists have forecast the possibility that the sub-tropical belt will migrate northwards in the next few decades, so that the northern boundary reaches a point beyond the central divide of the Qinling mountains or even the Huanghe. In addition, most of the frozen earth in the north-east and Qinghai-Tibet plateau and smaller glaciers in the Qilian and Tianshan mountains would thaw (Freeberne 1965a). Whereas summer rains would increase in Hainan, most of China's mid-latitude areas and the dry north-west would have less rain in the next century. Also, anticipated sea level rises would affect coastal areas from Hainan to Liaoning; some land would be submerged, tracts of saline soil would increase, drainage of waterlogged fields and coastal naviga-tion would become more difficult, and ideally there would have to be major investment in constructing dykes and seaports (*XNA*, 4 February 1989).

Parts of lowland China's climate are classified as humid. Truly mighty rivers descend from the uplands to the plains as well. Yet according to one definition, an optimum population based on the renewable fresh water resource has been set at 250 million. With the expansion of population, the problem of water supply has become much worse in the past forty years. Domestic users must compete with both industry and agriculture for a scarce commodity. Over 200 major cities do not have enough water, and a quarter of these face acute shortages. The situation is especially serious in the agriculturally vital north China plain and the municipalities of Tianjin and Beijing. At current consumption levels the demand for water in the capital alone is expected to increase by 50 per cent in the next few years (*Hongkong Standard* 18 November 1989; hereafter *HKST*).

Radical, even grandiose solutions which have been proposed include the long-distance inter-basin transfer of water from the Changjiang northwards, via three main routes. Work is well advanced in the most easily engineered region to the east, as improvements in the Haihe basin reveal (Freeberne 1972). Given the shortages, it is imperative that existing sources are not degraded by pollution. Yet Beijing itself has no sewage treatment plants, for example, and according to the chairman of the National Environ-ment Protection Agency, Qu Geping, there is no money to rectify the situation. Professor Zhang Gangdou, of Qinghua University, recently even claimed that the metropolitan water crisis was now simply a matter of life and death, and insisted on the immediate imposition of strict population curbs. Meanwhile, nationwide, domestic and indus-trial waste discharged untreated is destroying water resources. For example, in 1988 most of the 26.8 billion tons of industrial waste water was discharged raw, without any treatment whatsoever.

Any British visitor to China must brace themselves to be asked questions about London's pea-souper fogs, but in fact the tables have been turned.

Air pollution in China is so bad, that in some cities such as Chengde, Hebei and Chongqing, Sichuan, the concentration of total airborne particles and sulphur dioxide in the air has reached or exceeded the minimum values of those taken from the starting stage of the smog incident in London in the 1950s.

(*XNA* 18 February 1990)

Coal burning is the chief air polluter, and is blamed for the high incidence of tuberculosis. According to the United Nations, Shenyang, Liaoning, ranks as the world's second most polluted city. In marked contrast, in 1990, Weihai, Shandong, was honoured as the first 'national clean city'. For the ambition to industrialize at all costs presented the Chinese with a dilemma, because rapid, large-scale industrialization was the necessary concomitant of other major international industrial revolutions. But might not the Chinese otherwise have found their own special solution, as happened with the repeated attempts, starting with the Great Leap, to set up small-scale rural industries, thereby anticipating in part the concept of socialism with Chinese characteristics (Freeberne 1971)?

In February 1990 Ma Shijun from the Academy of Sciences asserted that during the 1980s, 'the contradiction between economy, resource and environment became the focus that restricts sustainable economic development in every country', but that this focus 'showed up more conspicuously in China due to the pressure of population growth' (*HKST* 13 February 1990). Today, China is paying heavily as a result, the cost of pollution to the nation running at an estimated 12 billion *yuan* (HK$19.2 billion) annually, with China actually spending from 9 billion to 10 billion *yuan*, or about 0.7 per cent of GNP each year on environmental protection. In fact, China aims to raise this figure to 1 per cent by 1992, which will still fall short of a recommended allocation of 1.5 per cent GNP.

## A Great Green Wall: the greening of China

Commonly, the unrelenting industry of the Chinese peasant, the heavy application of night soil and the presence of rich vegetable farming in the vicinity of major cities such as Shanghai and Guangzhou, give the impression of great natural fertility. Yet in the 1950s it was recognized that about 40 per cent of China's cultivated area comprised soils of poor quality. Changing the agricultural system in many areas, for political and social reasons, has led to inappropriate farming management and techniques, like extensive deep ploughing, and exposed again the true nature of environmental constraints. Despite new soil amelioration programmes, the overall difficulties which China's agronomists faced have intensified into the 1990s. In consequence, there is a significant dual problem of a contracting area in agriculture, coupled with the diminishing fertility of the area remaining.

Soil erosion has become menacing in the last forty years. In the 1950s 1.16 million square kilometres were affected; now the figure is 1.5 million, including roughly a third of the total cultivated area, with 51 million hectares of grassland threatened by desertification, or a quarter of the national pasture.

Officially only 13 per cent of China, or 124.65 million hectares, is in forest. The excessive felling of the residual natural forests, now largely restricted to the relatively remote and thinly settled northeast and south-west, and of secondary tree cover elsewhere has resulted in accelerating soil loss, as the great river systems carve through regions such as the friable loess plateau, disgorging huge amounts of silt into the sea every year. Afforestation schemes have met with mixed success, low survival rates for newly planted tracts being a perennial problem. But at least until the early 1980s, when agriculture remained collectivized, winter tree planting and soil improvement measures were attempted, whereas under the responsibility system – with its every-man-for-himself mentality – such programmes have been largely neglected, with some notable exceptions, described below. In any event, currently, 100 million cubic metres more timber is felled annually than reaches maturity. Also, forest fires, plant diseases and insect pests and illegal, uncontrolled logging are all serious problems, causing a longstanding shortage of timber products.

One of the most ambitious projects has involved the planting of a Great Green Wall, to hold back the advance of the deserts to the north. However, without extensive field access, and given earlier failures, it is difficult to assess the overall success of such a vast scheme. Stretching for 7,000 kilometres, and from 400 to 1,700 kilometres wide, the project is described as the biggest 'ecological endeavour' in the world, aimed at protecting 4 million square kilometres, or 42 per cent of the mainland, covering the desert regions and loess plateau. For three decades, desertification buried an average of 700,000 hectares a year, until the decision to plant the Great Green Wall was taken in 1978.

Beneficial results so far recorded include the marked reduction in Beijing's notorious duststorms: since 1985, down from about 30 dusty days in the 1970s to an average of 12.2 days annually. The story of China's greening does not end here, however. In 1955 coastal forests covered only 346,000 hectares. By 1988 they totalled 5.33 million hectares, hugging just 8,000 kilometres of the coast. In March 1989 a second giant, unified scheme to safeguard the entire 18,000 kilometre-long coast of mainland China with a protective forest skin was started, with 3.35 million hectares to be planted by 2010. Accounting for only 2.6 per cent of the surface area and 195 *xian* (counties), this coastal zone is demographically and economically supremely important with 10 per cent of the population, annually producing over 100 billion *yuan* in industrial and agricultural output.

Finally, attention has concentrated upon the main agricultural plains: in the north-east, north China, the middle and lower Changjiang and the Zhujiang River Delta – producing grains, cotton, vegetable oil and livestock – with 43 million hectares and 400 million people. Mostly long settled, the plains were mainly denuded of tree cover by 1949. Droughts, dust storms and floods conspired to reduce soil fertility drastically. Planting began in the 1950s, but it is only in the 1980s that work has started to be properly co-ordinated. By 1989 it was claimed that at national level, out of 908 *xian* situated on plains or semi-plain terrain, 250 had a forest cover of at least 10 per cent – a level which all 908 would reach by 2000; also, whereas the forest cover in the central north China plain was only 2 per cent in the 1950s, it was now 10.7 per cent; and overall 10 million hectares, or a quarter of the entire area of plains farmland, was protected by shelter belts.

Nature conservation has become something of a fad in China, as the country turns no matter how belatedly towards the preservation of its fauna and flora. Roughly 100 rare species are found only in China: amongst them the giant panda, golden monkey, white-lipped deer, Yangtze alligator, giant salamander, black-necked crane and brown-eared pheasant. The first nature reserves were established in the mid-1950s, but, as with so many aspects of development, by far the greatest progress has been made in the last ten years or so. By 1988 China had 481 'wilderness areas', extending over 20 million hectares, or 2.41 per cent of the surface.

There are three types of nature reserves. First, for forests and other forms of vegetation, particularly rare plant species, such as

Changbai (Jilin), Wuyi (Fujian) and Dinghu (Guangdong) – all three part of the United Nations Man and Biosphere network, and three of the six in China to be designated permanent world reserves. Second, wildlife reserves which seek to preserve both animals and the environment necessary for their very survival, of which Wolong, Sichuan, home of the giant panda is known world-wide. The panda is threatened by a peculiar combination of circumstances: by its seeming reluctance to mate, by the 1983 flowering and withering of bamboo and by poaching and smuggling. A recent, curious example of international co-operation involved the Milu (Père David's) deer, on the brink of extinction when, in 1985, Britain returned a number of deer raised on a private estate; the Chinese herd of about seventy lives on the Dafeng Milu Reserve in Jiangsu. Third, there are reserves of geomorphological character – for example, areas of karst scenery, or glaciers.

Nevertheless, many difficulties remain. For instance, some

officials put more emphasis on economic efficiency than they do on the preservation of nature. This is not a simple issue, since many people in remote areas, who are asked to make sacrifices for the sake of the environment, are themselves quite poor and in need of help to gain a decent living standard.

(*CT* February 1990)

# CHINA'S PLACE IN THE WORLD: THE SHAPING OF FOREIGN POLICY

Deeply ingrained prejudices have played an important part in determining China's foreign policy. So much so, that it is surely relevant to ask which policies stem from more recent expansionist doctrines of international socialism, and which from the distant past, particularly where irredentism, or, to use the jargon, Han chauvinism, has shown itself. For there have been episodes of naked nationalism over the last forty years, which partly negate the apparently more benign aspects of Chinese policies. This might be regarded as inevitable, given the pronounced ethnocentrism of China's long history; for this, after all, *is* traditionally the Middle Kingdom; the Chinese *are* the inheritors of the oldest and richest surviving civilization; and hence, by definition, all other countries are peripheral, and their peoples to a varying degree barbarian and inferior.

Since 1949, generally the evolution of China's relations with the outside world stalked the changes taking place in the both domestic politics and in the country's economic geography. For instance, when China was preoccupied by reconstruction and the successful launch of the First Five Year Plan, 1949 to 1957, whilst resolutely resisting the perceived threat from US imperialism, foreign policy was relatively low-key and based on the principles of peaceful coexistence. Then, all of a sudden, in 1958, the vaunting ambition of the Great Leap Forward followed hard on the heels of Mao Zedong's visit to Moscow in 1957, when he issued his famous dictum that the East Wind now prevailed over the West Wind. The economic set-backs caused by the Great Leap forced China to retreat diplomatically and concentrate on repairing the damage, 1959 to 1965. Also, in 1964, the Sino-Soviet rift became public, and can be viewed partly as fall-out from the period of extreme economic hardship, made worse by the withdrawal of Soviet technical assistance in 1960; had it not been for this betrayal the dispute might have remained under wraps, at least for longer.

Similarly, the essentially domestic yet near-crazed ructions of the Cultural Revolution, 1966 to 1976, spilt over into the international arena, particularly in the opening, florid phase, characterized by virulent xenophobia, 1966 to 1967 (Freeberne 1965b). China became an outcast, virtually friendless. By 1969, China was fighting a border war with the Soviet Union, and yet by 1972 received Nixon in Beijing. Both ping-pong and panda diplomacy were now in vogue, as in 1972 China belatedly took its seat in the United Nations. As part and parcel of these parallel workings, there was an actual intermeshing of political and economic factors. For example, in the Sino-Soviet dispute, apart from withdrawing technical aid from China, the Soviets were accused of earlier asset-stripping, by removing industrial plant from Manchuria immediately prior to Liberation and by taking an unfair share of strategic minerals from Xinjiang, when joint-stock companies were established in the north-west in the 1950s (Freeberne 1968). Likewise, the cooling off in relations resulted in a sharp decline in trade between the two socialist giants, just as grudging rapprochement brought about an upturn since the late 1980s.

Another potentially contentious territorial issue, concerns Chinese claims to off-shore islands. First, to the east, in the shape of the Diaoyutai, off the coast of Taiwan. Second, to the south, the Xisha and Nansha, extending as far south as 4 degrees north of the

Equator. Although the islands are tiny, they may have off-shore oil resources and, apart from the two Chinas, Vietnam, the Philippines and Japan remain interested parties. Whilst insisting on sovereignty, China is now prepared to discuss with relevant countries the joint development of islands in the South China Sea.

Mao Zedong's visit to Moscow in 1957 coincided with the fortieth anniversary of the October Revolution. On Mao's shopping list was a request for help in developing China's nuclear capability. The plea was turned down. The Chinese were intensely angry. Strategically the atomic bomb was dismissed as a paper tiger. Nevertheless, in 1958, Mao predicted that China would have both the atomic and hydrogen bombs within ten years. Mao's prediction proved correct, as China exploded an atomic device in 1964 and a hydrogen bomb in 1967. Also, the American backing for Taiwan stood in the way of any hopes of reunification of the two Chinas, coming to a head in the Taiwan Straights crisis in 1958.

In the opening decade, whilst apparently eschewing great power status, China nevertheless harboured ambitions towards assuming a leading role amongst, and acting as spokesman for Third World nations, by preaching the Five Principles of Peaceful Coexistence, particularly after the Bandung Conference in 1955: respect for each other's territorial integrity and sovereignty; non-aggression; non-interference in each other's internal affairs; equality and mutual benefit; and peaceful coexistence. Initially, the diplomatic offensive tended to focus on neighbouring states in Asia, not least India, then especially, often regarded as representing a rival economic model of development. However, this special relationship exploded with the Tibetan Uprising in 1959, after which, in the eyes of many, the Chinese lost considerable face and thereafter could never be wholly trusted, so that the diplomatic net had to be cast further afield in succeeding decades.

To illustrate: in 1964, Zhou Enlai declared that the African continent 'was ripe for revolution'. China went on the diplomatic offensive in developing political, economic and cultural links with many African nations, and by 1965 the Chinese were involved in at least a dozen countries. The construction of the Tanzam railway was a classic example of the way in which China sought to extend its sphere of influence in Africa by ostensibly making friendly overtures, and by supplying substantial financial, technical, material and man-power assistance. The project also indicated some of the pitfalls which can be experienced in establishing overseas connections. As often as not, the Chinese simply did not *get on with* the local population. The thin veneer of friendship all too frequently fragmented, due to the marked antipathy which many Chinese feel towards black peoples, which is even greater than that displayed towards other foreigners.

It would be easy either to exaggerate or to trivialize such racial differences. The fact is that they are the cause of ongoing tensions. For example, the sorry experiences of African students studying in China has been well documented. Late in 1988, there was trouble with African students in Nanjing, which then spread to a number of other centres of learning.

The post-1978 reforms quickened a trend towards positive diplomatic activity, where the commercial instincts of the Chinese have been encouraged by the open door policy and, in turn, have been influential in deciding the nature of China's links with the outside world. Thus, economic pragmatism may now be every bit as important as ideology in determining the future of external relations.

## Current international relations

For years, as far as the Chinese were concerned, three obstacles stood in the way of negotiations with the Soviet Union, which might otherwise have led to normalization in relations. First, Soviet troops had to be withdrawn from the long northern frontier. Second, the Soviet Union must leave Afghanistan. Finally, the Soviet Union must stop interfering in Vietnam and Kampuchea.

President Gorbachev flew to Beijing in May 1989, for the first Sino-Soviet summit for thirty years, only to find that what should have been an epoch-making event was completely overshadowed by the student democracy movement. The Moscow-trained, unbending technocrat, Li Peng – blamed by many for the Tiananmen episode – as prime minister made a return visit to Moscow a year later. There was little evidence, however, of a major breakthrough in relations between the two nations, such as a resolution of border disagreements, for example. Despite superficial appearances, the two were not natural allies. At best, links remained lukewarm; at worst, antipathetic.

In any case, both countries have become preoccupied by the recent sweeping changes in world socialism, resulting in domestic turmoil throughout the former Eastern bloc. As for China, a return to the civil wars of the 1920s and 1930s is to be avoided at all cost, and in fact there are some who see cause for hope in the June 1989 Tiananmen incident. During the Great Leap Forward of 1958, one reason for building the backyard iron and steel furnaces was to introduce the peasant masses to industrial technology, no matter how crude. Likewise, although the student democracy movement was largely spontaneous, nevertheless various sectors in Chinese society gained their first taste of democracy. In the mean-time, the Chinese dragon and the Soviet bear engaged in endless ideological shadow boxing, with China stealing the initiative in March 1991, by aiming both to assist and to humble its 'elder brother' with a commodity loan of 1 billion Swiss francs, thereby reversing the historic relationship. However, although the Chinese foreign minister, Qian Qichen, publicly hoped for stability and the success of *perestroika* in the Soviet Union, privately China fears *perestroika* and blames Gorbachev for the break-up of the Eastern bloc. Equally dramatic was the April 1991 announcement that the two socialist giants were close to signing an agreement on the demarcation and demilitarization of the entire 7,300-kilometre common border. Following the failed Soviet coup in August 1991, matters were necessarily placed on hold.

International political links were severed by Tiananmen; economic and technical aid were suspended both by individual countries and by international agencies, such as the World Bank; trade was severely hit; tourism all but dried-up and educational, scientific and cultural connections were disrupted. Mostly, these adverse results persisted beyond the first anniversary of Tiananmen.

Amazingly, however, the instinctive international boycott of China was breached by the secret American Snowcroft mission in late 1989, and by a British delegation, supposedly to do with the future of Hong Kong, in mid-1990. Foreign powers are every bit as likely to unashamedly seek out political and economic advantage in China at the end of the twentieth century, as was the Macartney mission in visiting the Chinese court at the end of the eighteenth century, in 1793. By early 1990 Japan was in the vanguard in working towards reopening trading and aid links with China, and continuing to develop the natural complementarity of the two economies.

Peaceful trading world-wide was placed in jeopardy by the outbreak of the Gulf War in mid-January 1991. To some, China's stance in the UN Security Council indicated a growing maturity following Tiananmen; whilst others detected a trade-off, whereby in return for Chinese acquiescence in its voting on Gulf resolutions the world would turn a blind eye to its continuing violation of human rights. Whatever its motivation, from the outset of the crisis China opposed Iraq's occupation of Kuwait and demanded unconditional withdrawal. At the onset of hostilities, China was adamant that the war must not be allowed to escalate and that all avenues be explored to bring about a peaceful solution; otherwise the probable destruction of oil fields would adversely affect the international economy and might bring further unrest to Muslim areas in China.

## Trade and foreign investment

The composition of Chinese trade has undergone major changes over the last four decades, with a heavy swing away from Eastern bloc to Western partners. At first, there was an emphasis on the export of primary produce and consumer goods and the import of manufactured goods and machinery. Then, as the quantity and range of China's industrial exports began to rise, increasingly the quality of the goods came under scrutiny. Throughout, the grain trade has remained a politically and economically sensitive issue.

During the Seventh Five Year Plan, 1986–90, China's foreign trade and investment in China expanded at a record rate, becoming ever more sophisticated. Trade volume totalled US$370 billion, an increase of 61 per cent over the previous plan; exports accounted for US$195 billion and imports for US$175 billion. In 1990 alone, exports topped US$62 billion. Finished industrial products made up 65.3 per cent of exports in 1989, compared with 45.8 per cent in 1985. Advanced technology, key equipment and materials needed for the domestic economy comprised 80 per cent of imports, 1986–90.

Approximately 22,000 overseas funded projects were approved during the Seventh Five Year Plan, with pledged investment of about US$23 billion, signifying considerable promise for China's economic development, especially as memories of Tiananmen faded.

Over the last decade, China has promoted labour export in order to earn hard currency, thereby partially reproducing the pre-Liberation history of indentured coolie labour. At least 350,000 Chinese have worked on about 8,000 Sino-foreign contracts worth US$12 billion. With a world labour market handling transactions worth about US$420 billion (1989) annually, Beijing aims to become 'an international giant' in supplying labour. Under a three-year plan, developing countries which receive aid from China were targeted, together with construction projects in the Soviet Union and Eastern Europe, and property developments in Thailand and the Americas. Chinese experts calculated that the Soviet Union alone needs 5 million workers in the short term and 50 million in the long term just to develop Siberia and the Soviet Far East.

## CONCLUSION: THE NEED FOR RESEARCH RECIPROCITY

The limitations of the individual geographer must rank high on the list of problems blocking a fuller appreciation of recent changes in the geography of China. As it is, regrettably few British geographers have chosen to specialize in this exacting, but immensely rewarding field, where even fewer tangle with the complex language (Freeberne 1985). Other barriers to research include: first, the problem of the very size of China. Second, China's physical, political and psychological remoteness. Third, the difficulty, if not the impossibility of gaining unfettered field access to investigate any subject other than limited physical topics; themes in human geography, which demand detailed examination, may be proscribed internally. Next, once in China, conventional entrée into and use of the unconventional libraries and archives requires dogged persistence, infinite patience and a large helping of luck; map collections are normally restricted access only. Finally, when located, all Chinese data must be handled with extreme caution.

Indeed, keeping matters in some sort of perspective is much to the point in China research, where there is always the temptation to make out a special case. For the question constantly nags: is it the distinctiveness and the differences, or is it the similarities and the bonds of common humanity, which exert such strong attraction? Without a doubt, treading the interface between politics, which has *determined* such profound and rapid changes in the geography of China, and geography provides quite one of the most fascinating and demanding areas of research.

Perhaps the Chinese place too much stress on the *direct* economic benefit to be derived from research, when it comes to approving a project. Certainly, a more imaginative approach might allow investigations into the structure, functioning and motivation of rural communities, for herein lies the heart-beat of China. If the immense productive forces represented by the peasant masses are to be truly liberated and encouraged to serve the state in the best sense of the term, by obtaining the highest standard of living for the greatest possible number of people – then how better than by reaching an intimate understanding of their *raison d'être*?

But herein lies the rub, for most cadres and intellectuals alike hold the countryside in contempt. That is why so many were rusticated during the Cultural Revolution: because they had lost touch with the grassroots. Could it be that, similarly, many in the present leadership are divorced from reality, despite the warning given by Vice-President Wang Zhen in early 1991, that the rural Party powerbase was being eroded, as the villages were being being overrun by feudalistic forces of exotic religion, capitalism and the clans, supposedly eradicated by the Chinese revolution? Sending out several battalions of social scientists into the rural areas might act as a relatively painless corrective.

An International Geographical Union (IGU) meeting in Beijing in August 1990 claimed considerable progress over the last decade in a range of fields: land use; agricultural resources; desertification; global climatic change; resource and territorial management; regional planning; remote sensing; geographic information systems; urban geography and historical geography. Also, at an international seminar in Shanghai in November 1990, experts agreed that for thirty years at least human geography had been neglected seriously at the expense of physical geography.

Within the cultural complex alone, there is an almost unending list of possible research challenges, all more or less urgent. The significance of race, language, religion and nationalism has been shown in several references to the national minorities. Yet such questions do not rest with the simple division between Chinese and non-Chinese, but must be extended to include the differences between Chinese and Chinese. If it is true that the Cantonese are a race apart, what about the Hainanese, happy as often as not to cock a snook at the mainlanders, now since 1988 masters of their own province and the largest of the SEZs? Undoubtedly controversial ground, difficult to investigate, but no less important for all that.

Despite language reform, illiteracy levels remain disastrously high at over 180 million (15.88 per cent) in 1990. From the 1950s onwards, under language reform *guoyu* was promoted as the national language, to take precedence over the many local and fragmented dialects, which show changes within the space of a few kilometres in Guangdong, for example, but where Cantonese still remains the everyday language in the villages, and commonly in the towns, too. Also, the written language was simplified under language reform. Flying in the face of Chinese tradition, education has been seriously underrated, undervalued and underfunded. During the Cultural Revolution, many children had no regular schooling for up to four years. Under the post-1978 economic reforms, as wealth has been generated, more and more children have been required to help with the family-based responsibility system, so that children have started to leave school early, and truancy is a growing problem. These, and related issues, await further research.

In theory at least, religion should be a non-starter where research is concerned. Whereas the constitution permits freedom of religious practice, communism decrees that all religion is mere superstition, which will die away as the excellence of the socialist system reveals itself. Also, there has been active suppression of religion since 1949, especially during the Cultural Revolution and in the areas occupied by national minorities.

Notwithstanding this background, as personal field work amongst the Deng clan in Hong Kong and Guangdong has revealed, old beliefs have proved remarkably resilient. In particular, a royal ancestral tomb was reconsecrated in China in 1988, with money from the New Territories lineage. In 1981 and 1988–9 the author was invited to China as a Senior Advanced Scholar, to conduct research approved by the State Education Commission in Beijing and supported by the British Council. However, upon arrival he was not allowed to visit either the Deng village at Shijing, Guangdong – the site of the grave of the Song dynasty princess – or Baisha, the original Deng village in Jiangxi. When research funding is scarce in both countries, this seems a sad waste of resources. Very few human geographers from overseas have been permitted to undertake field work in China since 1949, although a growing number of Chinese social scientists are able to research freely in foreign countries. Until there is true reciprocity, it will be possible to snatch only fleeting insights into current changes occurring in China, which have such profound implications not merely for a great country but also for the global village.

## REFERENCES AND BIBLIOGRAPHY

Banister, J. (1987) *China's Changing Population*, Stanford, Calif.: Stanford University Press.
Blunden, C. and Elvin, M. (1983) *Cultural Atlas of China*, Oxford: Phaidon.

Buchanan, K.M. (1970) *The Transformation of the Chinese Earth*, London: G. Bell.

Cannon, T. and Jenkins, A. (eds) (1990) *The Geography of Contemporary China*, London: Routledge.

China Financial and Economic Publishing House (1988) *New China's Population*, London: Macmillan.

Clayre, A. (1984) *The Heart of the Dragon*, London: Collins/Harvill.

Cole, J.P. (1985) *China 1950–2000: Performance and Prospects*, Nottingham: Department of Geography, Nottingham University.

Elvin, M. (1973) *The Pattern of the Chinese Past*, London: Methuen.

Endicott, S. (1988) *Red Earth: Revolution in a Sichuan Village*, London: I.B. Tauris.

Freeberne, M. (1962) 'Natural calamities in China, 1949–61', *Pacific Viewpoint* 3(2): 33–72.

Freeberne, M. (1964) 'Birth control in China', *Population Studies* 18(1): 5–16.

Freeberne, M. (1965a) 'Glacial meltwater resources in China', *Geographical Journal* 131(1): 57–60.

Freeberne, M. (1965b) 'Racial issues and the Sino-Soviet dispute', *Asian Survey* 5(8): 408–16.

Freeberne, M. (1966) 'Demographic and economic changes in the Sinkiang Uighur Autonomous Region', *Population Studies* 20(1): 103–24.

Freeberne, M. (1968) 'Minority unrest and Sino-Soviet rivalry in Sinkiang', in Fisher, C.A. (ed.) *Essays in Political Geography*, London: Methuen: 177–209.

Freeberne, M. (1971) 'China promotes local industries', *Geographical Magazine* 43(8): 505–11.

Freeberne, M. (1972) 'Haiho river basin project', *Geography* 57(3): 217–25.

Freeberne, M. (1985) 'Teaching China: a link group for geography', *Geographical Journal* 151(1): 145.

Freeberne, M. (1988) 'The 1987 Heilongjiang forest fire', unpublished manuscript.

Freeberne, M. (1991) 'A reconstruction of the 1991 floods in China', unpublished manuscript.

Freeberne, M. and Fung, K.I. (1981) 'Shanghai', in Pacione, M. (ed.) *Problems and Planning in Third World Cities*, London: Croom Helm.

Gittings, J. (1989) *China Changes Face: Socialism and Reform 1949–1989*, Hong Kong: Oxford University Press.

Glaeser, B. (ed.) (1987) *Learning from China: Development and Environment in Third World Countries*, London: Allen & Unwin.

Goodman, D.S.G. (ed.) (1989) *China's Regional Development*, London: Routledge.

Goodman, D.S.G. and Segal, G. (eds) (1989) *China at Forty: Mid-life Crisis?*, Oxford: Clarendon Press.

Huang, P.C.C. (1990) *The Peasant Family and Rural Development in the Yangzi Delta, 1350–1988*, Stanford, Calif.: Stanford University Press.

Jones, P. and Kevill, S. (1985) *China and the Soviet Union, 1949–84*, London: Longman.

Kane, P. (1988) *Famine in China 1959–61: Demographic and Social Implications*, London: Macmillan.

Kirkby, R.J.R. (1985) *Urbanisation in China: Town and Country in a Developing Economy 1949–2000 A.D.*, London: Croom Helm.

Leeming, F. (1985) *Rural China Today*, London: Longman.

Linge, G.J.R. and Forbes, D.K. (eds) (1990) *China's Spatial Economy: Recent Developments and Reforms*, Hong Kong: Oxford University Press.

March, A.L. (1974) *The Idea of China*, Newton Abbot: David & Charles.

Murphey, R. (1980) *The Fading of the Maoist Vision*, New York: Methuen.

Myrdal, J. (1965) *Report from a Chinese Village*, London: Heinemann.

Pannell, C.W. and Ma, L.J.C. (1983) *China: The Geography of Development and Modernisation*, London: Edward Arnold.

Population Census Office of the State Council (1987) *The Population Atlas of China*, Hong Kong: Oxford University Press.

Richardson, S.D. (1990) *Forests and Forestry in China: Changing Patterns of Resource Development*, Washington DC: Island Press.

Ross, L. (1988) *Environmental Policy in China*, Bloomington, Ind.: Indiana University Press.

Sivin, N., Wood, F., Brooke, P. and Ronan, C. (eds) (1989) *The Contemporary Atlas of China*, Sydney: Collins.

Smil, V. (1984) *The Bad Earth*, London: Zed Press.

Sun, J. (ed.) (1988) *The Economic Geography of China*, Hong Kong: Oxford University Press.

Tuan, Y.F. (1970) *China*, London: Longman.

Whitney, J.B.R. (1970) *China: Area, Administration and Nation Building*, Chicago: Department of Geography, University of Chicago.

Zhao, S. (1986) *Physical Geography of China*, Beijing: Science Press.

# 6

# THE CHANGING GEOGRAPHY OF TAIWAN, HONG KONG AND MACAU

## Richard Louis Edmonds

## INTRODUCTION

During the last quarter century Taiwan, Hong Kong and Macau have been amongst the most dynamic economic entities in the world. Taiwan and Hong Kong, along with South Korea and Singapore have become known as the 'four little dragons' of Eastern Asia. Macau has not been included within this select grouping presumably because of its small size. Economic growth, however, has brought with it geographical and political problems which will plague Taiwan, Hong Kong, and Macau for many years to come. In this chapter we shall look at each of these polities separately and try to analyse the changes which they have been experiencing since the 1960s.

## TAIWAN, THE REPUBLIC OF CHINA

The territory controlled today by the Republic of China amounts to 35,981 sq km. Taiwan was a colony of Japan from 1895 to 1945. Since 1949, Taiwan has remained under Nationalist (Kuomintang) control along with the off-shore islands of Chin-men (Kinmen) and Ma-tsu (Lien-chiang County) in Fujian Province. Chin-men and Lien-chiang County are to end their period of direct military rule and to elect their first county magistrates in 1993. In addition, the Nationalists maintain garrisons and weather stations in the South China Sea on T'ai-p'ing Island in the Nansha or Spratly Islands and in the Tungsha or Pratas Islands (see Figure 6.1). The government still refers to itself as the Republic of China, although the majority of nations now recognize Beijing. Technically only the island of

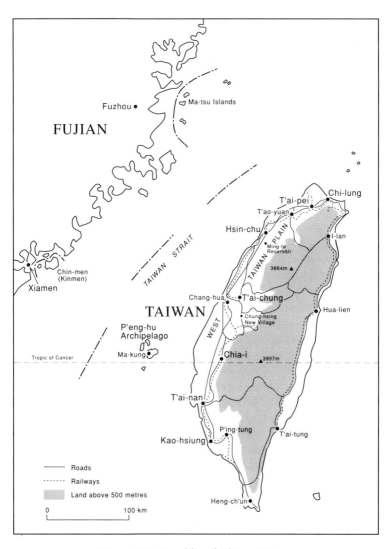

*Figure 6.1*   Republic of China: Taiwan

Taiwan, the P'eng-hu Archipelago, sixty-four islands to Taiwan Island's west, and twenty-one other small islands close to Taiwan Island are considered Taiwan Province. The cities of T'ai-pei and Kao-hsiung were separated from Taiwan Province in 1968 and 1979 respectively and are administered as 'directly controlled' cities.

161

T'ai-pei has become the temporary capital of the Republic while the Taiwan provincial capital has been re-established at Chung-hsing New Village in the centre of the island. In this chapter the terms Republic of China and Taiwan will be used more or less interchangeably, although Taiwan Province will be used to refer to that specific administrative division.

Both the mainland People's Republic of China and the Republic of China on Taiwan agree officially that there is only one China. Both reject any move towards a declaration of the independence of Taiwan, yet there is an unofficial independence movement in Taiwan, supported by independence organizations in Japan and the United States. The election results of December 1991 suggest that only a minority of people in Taiwan support independence at this time.

## Population growth

The population of Taiwan in December 1990 was 20,397,388 and, as Figure 6.2 shows, this represents close to a doubling of the 1960 population and a growth slightly over 20 per cent since 1980. The population density is over 564 persons per square kilometre, the highest density of any Chinese province and the second highest of any comparable world polity, after Bangladesh. As 64 per cent of the island is forested and mountainous and only 25 per cent is arable, there is an extremely high population density in the lowlands. The annual population growth rate has been dropping: going below 3 per cent per annum in 1966 and below 2 per cent per annum since 1977.

Despite a tendency for people to move towards the north and into cities, the geographical distribution has remained largely the same over the past thirty years with greatest concentrations in the northern portion of the West Taiwan Plain (over 2,000 persons per sq km) and lowest densities (20 persons per sq km) in the east central mountain range. The northern portion of Taiwan has 42.67 per cent of the population crowded into 10.2 per cent of the total area (Qín Kěxǐ 1989: 35; 'Běi Shì rénkǒu 1991: 7).

## Economic change

### Urbanization and communications

Urbanization has been rapid. Between 1965 and 1988, the proportion of the population engaged in agriculture dropped from 45.5 per

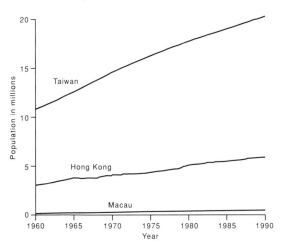

*Figure 6.2* Population growth trends in Taiwan, Hong Kong and Macau

cent to less than 20 per cent. T'ai-pei's population grew from 1.2 million in 1967 to 2.5 million in 1985. As of 1990 T'ai-pei had a population density of 10,160 per sq km which made it the most densely populated major city in the world. T'ai-pei has roughly twice the population of Kao-hsiung, which has approximately twice the population of T'ai-chung and of T'ai-nan.

Rapid urbanization has brought with it serious transportation and environmental problems which are now finally being addressed. In 1987 plans were mapped out for all of T'ai-pei and much of Kao-hsiung. In addition, the government is undertaking some new town development and has been building public housing since 1975 (Research, Development and Evaluation Commission 1988: 108). However, public housing appears to be too little too late as average floor space in Taiwan cities remains below that of Japan.

Urbanization has also brought about rapid changes in Taiwan's countryside as infrastructural development, factories, stock-brokers, and suburbanization have changed the rural lifestyle beyond re-cognition. Since many of those now in their 40s and 50s out-migrated to the cities in the 1960s and 1970s, a cultural gap has developed between the rural grandparents and their urban grand-children with the parents caught in between two worlds (Huáng Chūnmíng 1990: 4).

In 1974 the Nationalist Chinese government launched the 'Ten Major Projects' programme which finished in 1979. Of these ten projects, six were transport-related. This demonstrates how seriously the government has taken transport development. However, in T'ai-pei the need for public transport development has been critical. T'ai-pei has chronic congestion, no underground development until the late 1980s and as a consequence a confusing and congested bus network. A mass transit network is due to be completed by 1999 (Research, Development and Evaluation Commission 1988: 81). There is also a plan to construct a new high-speed railway between T'ai-pei and Kao-hsiung with a total of seven stations. The importance of future urbanization and communications development can be seen from the fact that urban development-housing and transport-communications are the two largest expenditure categories in the Sixth Five Year Plan beginning in 1991, accounting for 43.07 per cent of total investment.

One can already move from the northern part of the island to the south-west by dual carriage motorway. This represents a phenomenal change from the two-lane provincial roads with water buffaloes walking along the shoulders in the mid-1960s. The government will also begin to upgrade east coast transport during the 1990s by double tracking the 82-kilometre railway between I-lan and Hua-lien. There is also a plan to upgrade over 511 km of roads along the west coast into a dual carriage motorway and improve twelve major east–west roads with the latter project due to be completed in 1996 at an estimated cost of NT$116,840 million.

## Industrialization

The industrialization of Taiwan over the past three decades has been phenomenal. Since the 1950s investment has been directed at segments of industry which have high added value, particularly the manufacture of electronic goods. In 1966 the government took a further major step with the establishment of the Kao-hsiung Export Processing Zone, followed by two others at Nan-tze and T'ai-chung in 1969. These zones stimulated industrial exports, with emphasis on electronic goods. By 1980, the Chinese government was developing a different kind of zone, the Hsin-chu Industrial Park which has tried to attract more state-of-the-art investments both domestically and internationally. It has been said that since 1980 Taiwan has entered a 'technology-sensitive phase' of development (Li 1988: 28). Throughout, the government has controlled exchange rate policies and financial markets through its control of banks.

The role of private enterprise in the economy grew steadily between 1960 and 1980 from a share of 52.1 per cent to 80.2 per cent (Wu Qi 1988: 62). Like Hong Kong and Macau and in contrast to South Korea and Japan, much of Taiwan's industrial development has been undertaken by small firms. The major worry for the 1990s is the shortage of domestic investment which is in part due to the desire for entrepreneurs to emigrate or to take advantage of cheaper labour in places such as mainland China or Thailand. The goal of the government is to move the economy towards skilled labour-export orientated light industry by making Taiwan the leading supplier of information technology products by AD 2000. The Executive Yüan has proclaimed 'ten new flourishing industries' and 'eight key technologies' for the twenty-first century which are to be allocated considerable funds in the Sixth Five Year Plan beginning in 1991. The ten industries are: information technology, communications, consumer-orientated electronic products, precision machinery and automated enterprises, high-quality raw materials, semi-conductors, special chemical and drug enterprises, aviation, medicine, and pollution abatement ('Hángtài' 1990: 2).

### Agricultural change

Agriculture has been subsidizing industrial growth in Taiwan since the 1950s. While industrial development has been spectacular, agricultural development has been slow since the late 1960s and the importance of agriculture in Taiwan's economy has slipped considerably. Paddy, which is widely dispersed throughout the island, remains the most important crop, although its importance has diminished from 50 per cent of crop production in 1965 to slightly over 30 per cent in 1985 (Rèn Wéixìn 1990: 32–3). However, with the change over to a preference for more American-style foods, rice surpluses have been accumulating since the mid-1970s (Williams 1988: 31–2). Sugar cane production is concentrated in the south-west and tea is grown mostly on the red soil hill slopes of the northern and central parts of Taiwan. Vegetables and fruits have been making up a larger proportion of crop production since the mid-1960s as increased wealth has encouraged a greater variety of food stuffs. From the mid-1960s the total cropped area, the multiple cropping index and the area of paddy have all declined. Since 1965 Taiwan has not been self-sufficient in foods either in monetary value or calories and import dependency has been particularly heavy in terms of cereals, pulses, nuts and milk.

The farm population decreased from 5.74 million to 3.78 million between 1965 and 1988 with the income of farmers now considerably below average levels. An increasing number of farm households have had to turn to other occupations to make ends meet. Whereas on-farm income accounted for 87.2 per cent of total farm household income in 1965 this figure had dropped to 36.6 per cent in 1984 (Williams 1988: 29).

All these changes have been accompanied by a transformation of Taiwan's balance of agricultural trade from a US$105.9 million surplus in 1965 to a US$1,273 million deficit in 1985 with the turning point coming in 1973. Part of the deficit is due to a sharp rise in the increase in the import of feed grains (Council of Agriculture 1986a: 19–20). Agriculture's share in the net domestic product has dropped from 27.4 per cent in 1965 to 6.1 per cent in 1988. Although Taiwan's general economic growth has been strong, the increasing agricultural trade imbalance, declining food self-sufficiency, decline in numbers of full-time farmers and dependence on the USA as an agricultural trade partner pose serious problems for the future. In the future the government wishes to reduce food reserves to four months, loosen restrictions on agricultural land use, gear agriculture more to the domestic market than for exports, and reduce pig and fishery production in order to reduce pollution problems ('Nóngyè shēngchǎn' 1990: 7). One example of future trends was the conversion of the Da-chia Model Farm into Taiwan's first tourist leisure farm in the spring of 1991 ('Shǒuzuò xiūxián nóngyè nóngchǎng' 1991: 7).

## Environmental issues

While Taiwan's economy has grown rapidly since the 1960s, environmental consciousness, information and infrastructure have developed more slowly and unevenly. This situation has been particularly damaging as Taiwan made its name in plastics, petrochemicals, leather goods, pesticides and other high polluting industries. These industries were attracted to Taiwan as environmental consciousness was raised in her major markets – the United States and Japan.

As the Republic of China is about to enter the ranks of industrialized polities, the island is confronting environmental problems of an ever increasing scale. As an example the Chinese Environmental Protection Agency noted that the industrial city of San-ch'ung next to T'ai-pei had 170 days of bad air quality and only three days of good air quality in 1990 which was a decrease from four good days in 1989! (Táiběi Shì 1991: 7). Foreign firms are designing cosmetics which protect the skin against air pollution especially for use in Taiwan.

Social awareness of environmental issues and discontent with government and corporate management grew rapidly in the 1980s. The government has yet to come to grips with this problem and will have to give it top priority in the 1990s.

## *Resource depletion*

Taiwan was never rich in mineral resources and is increasingly resource deficient – meeting less than 8 per cent of its energy needs from domestic sources in 1988, which represents a significant drop from close to 67 per cent in 1965 (Council for Economic Planning and Development 1989: 98). Although some coal and natural gas reserves are found largely in the western half of the island, imported coal and petroleum have fired most of Taiwan's economic growth. Nuclear power has played an increasing role in the energy picture since its inception in 1977. In early 1991 water resources were becoming strained with reservoirs low and well-drilling companies experiencing a boom while the air force undertook cloud seeding. At the same time, there was considerable talk about limiting electricity supply in order to save energy.

Nature conservation in Taiwan has really developed only during the past quarter century even though the forestry law of the Republic of China was put into effect as early as 1945. The Taiwan Provincial Forestry Bureau established six natural protected areas and nature reserves between 1969 and 1989 and was the pioneer in conservation on the island as well as responsible for much loss of natural habitat (The Steering Committee Taiwan 2000 Study 1989: 116, 120). Four national parks have been established since 1982 covering 226,677 ha. Since 1984 the Council of Agriculture, has sponsored many ecological research projects (Council of Agriculture 1986b: 160–7). The Ministry of the Interior has also designated eleven coastal areas for protection. Although the distribution of potential and existing reserves are scattered, there is a shortage in the north-west and south-west.

Despite these developments, the Republic of China does not have any system for designating nature reserves nor any criteria for evaluating protection priorities (The Steering Committee Taiwan 2000 Study 1989: 116). Nature conservation law enforcement has not been as vigorous as it should be. While staff in the national parks have been working hard to do research, educate and prohibit exploitation, the national park areas are still not fully protected as exploitation of resources and private lands within the parks continues.

## Land degradation

The land surface of Taiwan shows the scars of heavy exploitation over the last thirty years. There were substantial drops in the total area of broadleaf forest, grassland and paddy between 1956 and 1977 and substantial rises in the amount of dry crop land, water area, mixed forest area, and urban and industrial land. The decrease in paddy is not offset by the increase in dry crop land as much of the best paddy went into urban and industrial uses. I was shocked to ride the train from T'ai-pei to T'ai-chung in 1989 for the first time since 1970 and to see the amount of derelict and improperly developed land along what used to be a scenic route. As forest lands have been logged or converted to agriculture often on steep slopes, soil erosion and siltation of reservoirs has increased. As an example, the Ming-te Reservoir, which was finished in 1970 and was expected to have over a fifty-year life, is already 20 per cent full of silt and 150,000 cubic metres of mud will have to be pumped out of the reservoir between 1991 and 1995 (Huáng Bìxiá 1991: 7).

Taiwan's slopes are prone to landslides and disasters have been common in the past. Locally there have also been problems caused by excessive mining of construction materials and coal. Rapid road construction during the last three decades means that only some of the highest mountain areas in the east central parts of Taiwan now remain undisturbed by human activity.

## Pollution

The rise in cancer deaths from fifty-seven to seventy-five deaths per hundred thousand people between 1961 and 1984 and the rise in congenital deformation of infants suggest that pollution has already begun to take its toll (The Steering Committee Taiwan 2000 Study

1989: 18). Taiwan has serious air, water, soil and noise pollution problems as well as solid waste problems. While levels of particulates in the air have been falling since 1969, the scant evidence available for other air pollutants, such as carbon monoxide, suggests that little improvement has been made in recent years. Water pollution is spurred on by high application rates for pesticides and fertilizers. Eutrophication is serious in all but three of Taiwan's rivers. Problems of heavy metals, pesticides, and polychlorinated biphenyls (PCBs) in Taiwan's soils have all appeared in the last thirty years. As of the late 1980s there was no systematic surveying system for noise measurement and laws for noise control date only from 1983–4 (The Steering Committee Taiwan 2000 Study 1989: 225). Noise from traffic and industrial machinery is serious. In all Taiwan less than 1 per cent of the human excrement receives primary sewage treatment. The figure for solid waste generated per capita rose annually from 1980 to 1985 with the vast majority dumped in landfills (The Steering Committee Taiwan 2000 Study 1989: 176).

The government estimates that during the 1990s there will be a need for an additional 200,000 environmental workers and investment should be over NT$1,000 billion. As of 1990 the Republic of China has a total of only 7,266 people working on environmental problems within government and the private sector. Of the mass of money to be spent during the 1990s, sewer construction and abatement facilities are slated to get the most funds ('Huánbǎofángwū' 1990: 7).

## Cultural context

### Religious affiliations

The religious affiliations of Taiwan are essentially the same as that found amongst Chinese communities elsewhere with a mixture of Confucian values and Buddhism, Dàoism (Taoism), ancestor worship and Christianity blended to varying degrees. In terms of formal numbers, Buddhists are the largest group claiming over 3.5 million souls followed by Dàoists numbering about 2 million and Christians about 0.75 million – the majority Protestant. In general, however, religion is secondary to secular activities.

The Nationalists' open stance on religion *vis-à-vis* the Communists has had some interesting implications. The Vatican still recognizes T'ai-pei as the official government of China. Saudi Arabia also maintained diplomatic relations with T'ai-pei until the end of the 1980s in part for religious reasons. The Nationalists have also stressed their respect for Confucianism, particularly when the Communists were engaging in anti-Confucian movements.

## Ethnicity, language and nationalism

The population of Taiwan can be divided into three major groupings: the Taiwan aborigines or Kaoshan, the Taiwanese, and the mainlanders. The aborigines are descendants of Austronesians who now number around 320,000 and are concentrated in mountainous central Taiwan. Their native languages do not belong to the Chinese language group. There are nine tribes: Atayal, Saisiat, Bunun, Tsou, Rukai, Ruyuma, Ami, and Yami (Government Information Office 1987: 39). Forced into the mountains by Han-Chinese settlers since the seventeenth century, the aborigines' role in Taiwan's affairs has become marginal and their traditional culture is fast vanishing. In early 1991 the aborigines put forward an effort to create their own autonomous region within Taiwan which was vetoed by the Interior Ministry.

The Taiwanese make up the majority of the island's population and are descendants of Han-Chinese settlers who arrived prior to the Japanese take-over in 1895. These people speak largely Min but in certain areas Hakka, both of which belong to the Chinese language group. Almost all of the people in the rural farming areas are Taiwanese.

The mainlander group is composed of people who moved to Taiwan from the mainland in the late 1940s, their descendants, and others who have subsequently arrived from mainland China. They largely reside in the cities. While these people may be speakers of any Chinese language, they mostly are not speakers of Min or Hakka and communicate amongst each other and with the Taiwanese and aborigines in Standard Chinese.

Ethnic conflict in Taiwan has largely revolved around the relationship between the Taiwanese who have held the land and the mainlanders who have traditionally held political power. Problems were very serious in the late 1940s and the early 1950s when Taiwanese were excluded from participation in various areas of government and mainlanders were refused employment in Taiwanese-owned enterprises (Winckler and Greenhalgh 1988: 146–7). Land reform actually helped the Nationalists to eliminate the Taiwanese landlord power base in the countryside. Many Taiwanese feel ambivalent to the Nationalist government with many former land-lords and small businessmen going to Japan and the United States where they have taken pro-Taiwanese Independence stands. Many of the elite who have remained in Taiwan are members of the opposition Democratic Progressive Party (Minchintang) or in-dependent political candidates.

Over the years the Taiwanese–mainlander differences have be-come blurred. Even by the 1960s, mainlanders and Taiwanese were finding it convenient to use each other as fronts to get around economic and political blocks that membership in their group represented. Eventually the Nationalists have had to incorporate Taiwanese into the ruling structure. The most startling example is the current president, Lee Teng-hui, who is chairman of the Nationalist Party and a Taiwanese. Most people on both sides of the Taiwanese–mainlander gap see the future as best provided for by stability. Although almost all identify themselves as Chinese, many do not want unity with mainland China for economic reasons and also do not want to declare Taiwan independent for fear of attack. The situation can be summed up by the saying that, 'unity with the mainland is something you can talk about but not implement and Taiwanese independence is something you can implement but not talk about'. In the long run, however, it appears that Chinese identity will win out over any pragmatic desires for Taiwanese nationalism, although there will need to be more advantages for Taiwan from unity with the mainland before most whole-heartedly support it.

## Political change

### Form of government

The major trend in government since the 1960s has been the increase in Taiwanese and opposition representation in the national assembly. However, the assembly is still dominated by Nationalist party elements, many of whom are mainlanders without mandate since 1948.

The division of the island into Taiwan Province, T'ai-pei Municipality and Kao-hsiung Municipality has been criticized as out-dated due to the rapid urbanization of T'ai-pei County,

Kao-hsiung County and the T'ai-chung area. Changes advocated include T'ai-pei and Kao-hsiung Municipalities annexing the majority of their corresponding counties and the T'ai-chung area perhaps being redesignated as a directly administered municipality. The impact would be to rationalize some of the municipal management problems and to strengthen the hand of the central government and the Nationalist Party in controlling the cities. There is also a plan to divide Taiwan into eighteen 'life spheres' for planning purposes during the next Six Year Plan ('Quánguó huàfēn' 1990: 2). Six of these spheres will be urban (T'ai-pei, T'ao-yuan, Hsin-chu, T'ai-chung, T'ai-nan and Kao-hsiung) and the other twelve will be 'general spheres' with travel time to the urban spheres to be reduced to one hour.

### Ideology and external affiliation

Nationalist Chinese ideology has been rather thin since Chiang Kai-shek assumed power over most of China in 1928. The main doctrine is Sun Yat-sen's Three People's Principles or *San Min Chu-yi*. The three principles are nationalism, democracy and people's livelihood. After the Nationalist retreat to Taiwan in 1949, anti-communism became the basic tenet of the government, which swore to 'save the Chinese compatriots on the mainland from the yoke of communism'. However, the Three People's Principles, especially people's livelihood, shows some close affiliations with socialism and state intervention in the economy of Taiwan has always been strong. Some of these industry–government connections in Taiwan have another origin – for example, government control over the Taiwan Sugar Corporation began under Japanese colonial rule. Therefore, while espousing democracy and capitalism, the Nationalists have actually held tight control over elections and undertaken considerable government intervention in the economy.

The greatest single political change within Taiwan over the last quarter century was the establishment and toleration of a significant opposition party, the Democratic Progressive Party in 1986. This transformation was preceded by a period of transition beginning during the 1960s in which Taiwanese and educated technocrats were allowed to reach high positions within government and the Nationalist Party. This transformation was caused by the expansion of higher education within Taiwan, the international crisis which the Republic of China was facing as the UN and many governments recognized Beijing, and the transformation of Taiwan from an agricultural to an industrial economy (Li Cheng and White 1990: 3–6). Chiang Kai-shek's son, Chiang Ching-kuo, was instrumental in both the transformation of the Republic of China from a state controlled by a mainland-born military elite into a government managed by technocrats with considerable Taiwanese participation.

## International context

### External trade and aid

In the 1950s Taiwan employed import substitution of non-durable goods as its major foreign trading policy. Due to the weakness of the domestic market this policy was only partially successful and aid from the United States of America was needed to make up for trade deficits. American aid stopped in 1965. From the 1960s, Republic of China economic policy focused on raising exports of durable consumer goods although import substitution continued with some change in the types of protected domestic goods. Since then the total value of the Republic of China's foreign trade has increased by more than a hundredfold. The Republic of China has had a favourable balance of commodity trade every year since 1976. In the 1980s the focus has been on developing high technology export industries.

The trading partners of the Republic of China have remained more or less the same since the 1960s. Imports from Japan have dropped from 40 per cent of the total in 1968 to slightly under 30 per cent by 1988. The United States of America remains a clear second with the level of annual imports consistently around 25 per cent of the total. Imports from other countries are all below the 5 per cent level. In recent years trade with Asian neighbours such as Hong Kong and Singapore has shown a modest growth in the total proportion of imports (see Table 6.1).

Although the Republic of China has enjoyed a favourable balance of trade since 1968 with its largest trading partner, the USA, it has had a trade deficit with its second most important partner, Japan, every year since 1956. In the 1980s Taiwan's favourable balances of trade with Hong Kong, Singapore and the United Kingdom have increased (see Table 6.2). Recent trade deficits are largely with resource rich countries such as Saudi Arabia and Brazil.

*Table 6.1*  Percentage of imports from selected countries to Taiwan, Republic of China

| Period | Japan | USA | FR Germany | Hong Kong | Australia | Saudi Arabia | UK |
|--------|-------|------|-----------|-----------|-----------|--------------|------|
| 1968 | 40.0 | 26.5 | 4.0 | 1.4 | 2.1 | 0.3 | 1.8 |
| 1973 | 37.7 | 25.1 | 5.3 | 2.6 | 2.6 | 1.1 | 1.9 |
| 1978 | 33.4 | 21.5 | 3.7 | 1.4 | 2.9 | 5.9 | 2.2 |
| 1983 | 27.5 | 22.9 | 3.4 | 1.5 | 3.4 | 9.5 | 1.5 |
| 1988 | 29.8 | 26.2 | 4.3 | 3.9 | 2.7 | 2.5 | 2.2 |

*Source: The Free China Journal* (24 September 1990: 8; 27 September 1990: 8).

*Table 6.2*  Percentage of exports to selected countries from Taiwan, Republic of China

| Period | USA | Japan | Hong Kong | FR Germany | UK | Singapore | Canada |
|--------|------|-------|-----------|------------|------|-----------|--------|
| 1968 | 35.3 | 16.2 | 9.2 | 5.8 | 0.8 | 2.8 | 4.4 |
| 1973 | 37.4 | 18.4 | 6.6 | 4.8 | 2.5 | 2.9 | 3.8 |
| 1978 | 39.5 | 12.4 | 6.8 | 4.5 | 2.6 | 2.3 | 2.6 |
| 1983 | 45.1 | 9.9 | 6.6 | 3.4 | 2.5 | 2.8 | 2.9 |
| 1988 | 38.7 | 14.5 | 9.2 | 3.9 | 3.1 | 2.8 | 2.6 |

*Source: Taiwan Statistical Data Book* (1989); Republic of China Council for Economic Planning and Development.

## External association and alignment

In foreign affairs the Nationalists were closely aligned to the United States prior to their retreat to Taiwan and this attitude has continued despite the break in diplomatic relations in 1979. The most notable development since the late 1980s has been the increased warming in relations between the Republic of China and the former Soviet Union and their former allies in Eastern Europe. The reason for the warming is the desire for trade.

## Share of world wealth

From the mid-1960s to 1973 the Republic of China had a fairly balanced balance of payments. From 1973 to 1980 the balance fluctuated more drastically, presumably due to the oil shock. However, from 1980 to 1987 the Republic of China had a rapidly growing trade surplus and still has a considerable surplus. This has left Taiwan with some of largest cash reserves of any polity in the world.

Wages and savings in Taiwan have increased dramatically since the 1960s. In general, the Taiwan Chinese have used this wealth to accumulate capital for investment or for their children's education rather than to purchase large quantities of consumer goods. Increasingly Taiwan families are spending more of their income on leisure, education and transportation whereas the proportion spent on food and drink has decreased. Rent and utilities also consume a greater part of the average person's budget than three decades ago.

While wages have managed to remain quite competitive with countries such as the UK, the USA, Japan and France, the cost of labour has risen above that of mainland China, Thailand and other south-east Asian polities so that some of the labour-intensive, low-value industries have begun to leave. Since the mid-1980s, Taiwan Chinese have begun investing considerable amounts of money in mainland China.

## HONG KONG

Hong Kong is only one-thirty-fourth the size of Taiwan with a total area of 1,060 sq km. Yet the growing influence of this territory on the economy of the world has been equally important over the past quarter century and Hong Kong has played a crucial role in the increasing contacts between the People's Republic of China and the Western world. Hong Kong is politically divided into three portions; Hong Kong Island (76 sq km) which was ceded to Britain in 1842, the Kowloon Peninsula (10 sq km) which was ceded in 1852, and the New Territories which was leased to Britain in 1898 for ninety-nine years (see Figure 6.3).

### Population growth

Compared with Hong Kong's rate of population growth between 1950 and 1965 when the territory became the home of many refugees fleeing from China, the rate of population growth has slowed down (see Figure 6.2). However the population still has grown from 3.82 million in 1965 to 5.81 million in 1989 and estimates suggest it will be over 7 million in AD 2001.

Since the mid-1970s the natural birth rate and mortality rate have been approaching that of the developed nations and the population has been ageing. The immigrant nature of the population can still be

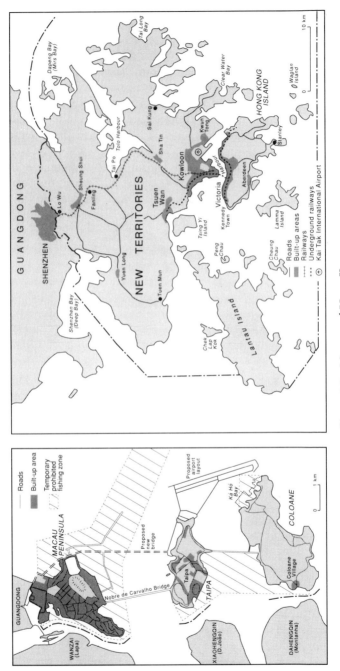

*Figure 6.3* Macau and Hong Kong

seen from the fact that only about 60 per cent of today's residents were born in Hong Kong.

Immigration from China and Vietnam was particularly heavy between 1978 and 1980 and again from Vietnam in 1988 and 1989. The slow-down from 1980 was due to the abandonment in that year of the 'touch-base' policy, which allowed illegal immigrants who reached urban Hong Kong to remain. The Vietnamese boat people have become a bone of contention in Hong Kong. Since June 1988 all refugees in Hong Kong must be screened to determine their status as genuine refugees and some of the refugees have been repatriated to Vietnam. Emigration from Hong Kong has increased since the signing of the Sino-British accord over Hong Kong's future in 1985 and particularly since the Tiananmen Massacre of 1989.

The population distribution in Hong Kong is uneven with densities of over 21,000 per square kilometre in the urban areas of Victoria on northern Hong Kong Island, the Kowloon peninsula, New Kowloon and Tsuen Wan while that in the majority of the New Territories is only 1,650 per square kilometre.

## Economic change

### Urbanization and communications

The most notable change in Hong Kong's urban pattern over the past three decades has been the growth of eight new towns in the New Territories which now are home to over one-third of the population. This huge programme, which was intended to relieve congestion in Victoria and Kowloon, cost the government HK$78 million since the late 1950s (Bristow 1989: vii). In addition there has been substantial suburbanization near Victoria and Kowloon since the early 1970s which has also helped to relieve congestion. New town development, however, has not been without its woes and Hong Kong still faces considerable housing shortages, commuting problems and over-concentration of people and industries.

Hong Kong's external trade is dominated by its harbour. Today it is the leading container port in the world in terms of throughput with the fastest turnaround time of any port in East Asia. This containerization was achieved without heavy reliance on Chinese cargo (Chiu and So 1986: 333–4). From 1949 up to the late 1970s the traditional entrepôt function was somewhat blocked by the animosity between the People's Republic of China and the West, particularly the United States. Since the beginning of China's open policy, the entrepôt trade has again become important with Hong Kong people taking the lead in trading and investing with the People's Republic. By the mid-1980s China was dominating the Hong Kong transhipment business.

Although the Hong Kong government has not been known to be bullish on public spending, about 50 per cent of its spending in the 1980s was on roads. Between the mid-1960s and 1990 the number of motor vehicles, particularly private cars, has risen tremendously. However, the vast majority of Hong Kong's residents are dependent on public transport. The public road transport is a mix of double-decker buses, trams (dating from 1904) in northern Hong Kong Island, and since the 1960s, public light buses (minibuses). Road transport received a great boost with the opening of the cross-harbour tunnel in 1972.

Railway transport finally received a boost with the opening of the Mass Transit Railway (MTR) underground network in 1979–80. The MTR system is still being expanded and a second Eastern Harbour Crossing for both MTR and road was completed in 1989. The MTR and the harbour tunnel projects involved both government and private enterprise. The Kowloon Canton Railway section in Hong Kong has expanded from 36 trains daily in 1966 to 474 in 1988, has been electrified and had the total number of stations increased although there has been virtually no new line development. A light railway was opened from Yuen Long to Tuen Mun in 1988 and this system is to be expanded in the early 1990s. Although ferry services are no longer the sole form of cross-harbour traffic they continue to play a role, especially in transport to outer islands.

Kai Tak International Airport has reached saturation point and plans to build a new airport north of Lantau Island with a bridge to Tsing Yi Island and the mainland were approved in 1989 with operation scheduled to begin in 1996. The airport plan became a point of political controversy between the governments of Hong Kong and the People's Republic of China with the Chinese not approving the project until July 1991. The Chinese main concern was that they do not want the British Hong Kong government to spend all their money on infrastructural development prior to 1997. The British have agreed to leave at least HK$25,000 million in reserves at the time of Chinese take-over.

## Industrialization

From the mid-1970s the manufacturing sector has showed very strong growth and despite a drop in numbers employed in recent years, still accounted for over one-third of total employment in 1988. In particular electronic goods and scientific equipment were expanding up to the mid-1980s. The growth of the clothing and textiles industries has, however, been slowing due to competition from other Asian states and the industry has begun to specialize in quality goods.

The constraint of space and need for industry to be geared to foreign markets encourages the scale of industries in Hong Kong to be small with the majority of companies employing fewer than 100 persons. It is necessary for any company to try and obtain a profit within a few years of its establishment. Thinking further than five years ahead is not the custom in Hong Kong business. Such flexibility is not present even in Taiwan, Singapore or South Korea.

Attempts at industrial zoning date to the 1950s but much of the industrial development has been haphazard with highly mixed land use common. Lack of space means that small industries have often been set up in domestic buildings and in squatter areas. These 'factories' are generally tolerated by Hong Kong's laissez-faire government. However, since the mid-1970s the government and private developers have been rehousing small-scale industries in flatted (multi-storey) factory estates.

Tertiary activities have declined in relative importance since the 1960s as manufacturing has grown. However, service industries are still the most important part of the economy. Tourism is Hong Kong's second largest industry after manufacturing. The Hong Kong Tourist Association was established in 1957 to promote the industry which has grown rapidly over the past thirty years. Visitors totalled 5.5 million in 1988 with those from Japan and Taiwan making up the largest proportions. In that year nine new hotels opened, which shows that financiers still feel there is further room for growth in the industry. However, the hotel industry as with so many of Hong Kong's industries is taking no chances and developing overseas operations in case they have to close down after 1997.

## Agricultural change

Only about 9 per cent of the land in Hong Kong can be considered as arable and 2 per cent of the population are engaged in agricultural production. Agriculture can no longer be said to play a major role in Hong Kong's economy, despite the fact that 38 per cent of the live poultry, 34 per cent of fresh vegetables, 18 per cent of live pigs, and 12 per cent of the fresh-water fish sold in Hong Kong are locally supplied. Better wages in industry and the availability of cheaper imports from China mean that fewer people are willing to undertake agriculture. Specialized high-priced products such as fresh vegetables, edible mushrooms and cut flowers now dominate local agriculture. Pond fish culture has increased from 186 hectares in 1954 to 1,400 hectares in 1988. Rice cultivation has dropped from 9,450 hectares in 1955 to less than one hectare in 1988 (Government Information Services 1989: 109). Even the area of land under vegetables and flowers has been decreasing since 1976.

## Environmental issues

Only since the 1970s has environmental protection become a priority of the government. The 1980s brought an increased commitment by the government to impose strict emission and noise controls, to ensure adequate collection, treatment and disposal of wastes, to upgrade older urban environments, and to require developers to undertake environmental impact assessments. However, the return of Hong Kong to China in 1997 leads to uncertainty in environmental planning. For example it may become easier to dump wastes in Guangdong Province or elsewhere than to look for sophisticated and expensive local solutions.

### Land degradation

With hills rising to over 900 metres it should come as no surprise that gullying and badland development can occur easily. Soil erosion problems are made more complex by the subtropical monsoon climate with hot wet summers and cool, dry winters which can cause rapid weathering.

During the 1960s and 1970s there were significant afforestation programmes and since 1976 the government has also been establishing country parks. However, transportation improvements and population growth mean that a larger number of industries, squatters and day visitors are using the formerly sparsely populated portions of the New Territories. As a consequence land degradation in rural

Hong Kong has been more serious in the last thirty years than in the preceding 120 years.

## Pollution

Urban development has brought about local climatic change, witnessed by a gradual warming of temperatures recorded at the Royal Observatory since 1884 (Chiu and So 1986: 103). In particular the night-time minimum temperatures in urban Hong Kong have increased with the rapid construction of tall buildings and the increase in anthropogenic heat released from buildings since the Second World War (Chiu and So 1986: 105).

Water pollution reached serious levels by the 1970s and got worse during the 1980s. The increase in the incidence rate of cancer since 1965 (*Hong Kong Monthly Digest of Statistics* March 1990: 110) may be one effect of the increasing pollution. As almost 90 per cent of Hong Kong's liquid effluent load is discharged directly (Chiu and So 1986: 368), a considerable amount of livestock manure, domestic waste, and toxic industrial wastes are polluting water bodies. Coastal pollution has increased especially in landlocked bays and in Victoria Harbour. From the late 1970s, toxic heavy metals have increasingly been found in various kinds of seafood harvested from the nearby coasts. Recently the quality of some of the most famous beaches has deteriorated and up to fourteen tonnes of floating refuse are pulled out of Victoria Harbour each day.

The government invested in a number of new sewage treatment plants during the 1980s. A Water Pollution Control Ordinance was passed in 1980 which gave various departments of the government more powers. Since then the government has made polluted areas into water control zones and has been requiring industrial and certain domestic effluent sources to obtain licences with allowable pollutant limits. The government also amended the Waste Disposal Ordinance in 1987 so that livestock keeping is prohibited in urban areas and livestock wastes in the rural New Territories will be progressively controlled (Government Information Services 1989: 314–15).

As wind circulation is good, pollutants are easily dispersed. Problems have been in low pockets surrounded by hills or in narrow streets with a number of tall buildings on either side. In the 1960s the major air pollution concerns were fumes from the Kennedy Town Incinerator and from power stations. Since then, more care has been taken to use low sulphur content fuels at the power stations on critical days leading to reductions in sulphur dioxide levels and smoke density levels. The Air Pollution Control Ordinance was enacted in 1983. In recent years, levels of sulphur dioxide, nitrogen dioxide and rainwater acidity have exceeded standards on occasions and particulates continue to exceed standards at virtually all monitoring stations.

The tremendous growth in economic activity has increased noise pollution since the 1960s. Kai Tak Airport has been a source of constant noise pollution which may disappear with the construction of a new airport. In 1988 a Noise Control Ordinance was passed. However, implementation of standards is still in the beginning stages.

From 1966, the government began to use incineration as a supplementary method to landfill tipping for disposal of solid wastes. Air pollution problems forced the government to revert to tipping by composting and high density baling (Chiu and So 1986: 375). Plans call for closure of the Kennedy Town Incinerator in 1992. There will be increased use of transfer stations with continued tipping at large landfill sites along with separate facilities to dispose of chemical wastes and composting for animal wastes. From 1990 it is planned that sludges will be dumped at sea to the east of Waglan Island.

With the nuclear power station at Daya Bay, about 50 kilometres to the north-east in Guangdong Province due to be completed in 1993, there is growing concern about the potential for radiation pollution. In anticipation, the government has stepped up its radiation background monitoring programme which first began in 1960.

## Cultural context

### Religious affiliations and linguistic variation

Hong Kong observes most of the common Chinese festivals, although the traditional customs of Guangdong Province dominate. Buddhism and Dàoism are strong, particularly amongst the older generation. There are also considerable Christian, Muslim, Hindu, Sikh and Jewish communities. A considerable number of the Christians and Muslims are ethnic Chinese.

The official languages in Hong Kong are English and Chinese, which in common practice means British English and Cantonese. In recent years the status of written Chinese has come closer to being on a parity with English although most government signs still put the English on top. With the take-over of Hong Kong by China imminent, there has been increasing interest in learning Standard Chinese.

## Ethnicity and nationalism

Those with origins in Hong Kong, Guangdong Province and Macau dominate Hong Kong's population. While most are native speakers of Cantonese, there is a significant number of Chaozhou speakers. The remainder of the Chinese community comes from all over China with the most prominent group coming from the Shanghai area. By contrast the British community is small. As a major international port, Hong Kong has significant minorities from all over the world.

The strength of Chinese ethnic identity and Hong Kong's small size has meant that nationalism has never taken root. Although there are many amongst the Chinese and British communities who are upset about the reversion of Hong Kong to the People's Republic, no one has suggested independence. Some Taiwan Chinese have suggested joint rule of Hong Kong by the Communists and Nationalists as a way forward for unity of the mainland and Taiwan. However, the People's Republic has shown no interest and the UK government has made no effort in this direction, presumably because it is against Beijing's wishes and could destabilize Hong Kong during the run up to 1997.

## Political change

### Form of government

A major change in Hong Kong occurred when the Joint Declaration of the Chinese and the UK government came into effect in 1985. From 1 July 1997 Hong Kong will become a Special Administrative Region of China.

Currently local authority is vested in the Governor, who has been given powers to run Hong Kong separately from the authority of London. The Governor is assisted by an Executive Council which from 1988 has included ten appointed members as well as four *ex-officio* members appointed by the Governor. In addition, there is a Legislative Council with a maximum membership of fifty-seven. It is expected that only ten of the Legislative Council members will be directly elected from 1991. This means that the Legislative Council is not a true democratic body. At the regional level there is an Urban Council serving the urban areas of Hong Kong Island and Kowloon and a Regional Council for the new towns and rural areas of the New Territories. Below these councils are nineteen district boards. Elections to these councils and boards on a geographical basis.

Since the Joint Declaration, Hong Kong has been acting more and more separately from the UK government and listening more to what Beijing has to say. This trend no doubt will continue.

## *Ideology and external affiliation*

Ideology for Hong Kong essentially is support of free enterprise capitalism in the extreme. Its philosophy is laissez-faire or minimal government involvement in the economy. Hong Kong has evolved into one of the world's major financial centres under this sort of policy as banking and currency exchange have had few controls. This philosophy has also meant that like Macau and in contrast to Taiwan, public spending has always been weak.

International relations are still largely handled through the United Kingdom government. However, since 1986, Hong Kong negotiates separately from the UK in GATT.

## International context

### *External trade*

Re-exports are increasingly important to the economy. In 1975 re-exports amounted to 20 per cent of total exports but by 1988 this figure had increased to 55 per cent with the greatest increase coming in the 1980s due to the growth of the China trade. Hong Kong has played a key role in trading goods between politically hostile states, primarily mainland China and other countries such as the United States, Japan, and more recently Taiwan and South Korea. Exports to Japan have been growing strongly in the 1980s (see Tables 6.3 and 6.4).

*Table 6.3*  Percentage of imports from selected countries to Hong Kong

| Period | PR China | Japan | Taiwan, R of China | USA | R of Korea (south) | Singapore | UK |
|--------|----------|-------|--------------------|-----|---------------------|-----------|-----|
| 1968 | 19.5 | 21.8 | 3.3 | 13.8 | 0.7 | 2.1 | 8.7 |
| 1973 | 19.4 | 20.2 | 5.8 | 12.8 | 2.1 | 3.3 | 5.9 |
| 1978 | 16.7 | 22.8 | 6.8 | 11.9 | 2.8 | 5.1 | 4.2 |
| 1983 | 24.4 | 23.0 | 7.1 | 10.9 | 2.9 | 6.0 | 4.2 |
| 1988 | 31.2 | 18.6 | 8.9 | 8.3 | 5.3 | 3.7 | 2.6 |

*Source:* Government Information Services (1968, 1974, 1979, 1981, 1985, 1989) Hong Kong.

*Table 6.4*  Percentage of domestic exports to selected countries from Hong Kong

| Period | USA | PR China | FR Germany | UK | Japan | Canada | Singapore |
|--------|-----|----------|------------|-----|-------|--------|-----------|
| 1968 | 41.4 | n.a. | 5.9 | 15.9 | 2.8 | 3.4 | 2.6 |
| 1973 | 35.0 | n.a. | 9.8 | 14.5 | 5.5 | 2.6 | 2.7 |
| 1978 | 37.2 | 0.2 | 10.9 | 9.5 | 4.6 | 3.1 | 2.7 |
| 1983 | 42.0 | 6.0 | 7.7 | 8.2 | 3.7 | 3.6 | 2.1 |
| 1988 | 33.5 | 17.5 | 7.4 | 7.1 | 5.3 | 2.7 | 2.4 |

*Source:* Government Information Services (1968, 1974, 1979, 1981, 1985, 1989) Hong Kong.

## External association and alignment

As trade dominates Hong Kong's life, it is necessary for the government to maintain as open a foreign policy as possible. This has been constrained in the past by UK foreign policy and in the future will be constrained by the foreign policy of the People's Republic of China. The current favourable political climate between the Commonwealth of Independent States and both the UK and China suggests that relationships with the former Soviet block should improve considerably.

## Share of world wealth

Hong Kong is a place of very rich and very poor people. In the overall picture, however, a tremendous amount of wealth has been concentrated in the territory although many of its wealthier citizens are in the process of emigrating prior to 1997. Hong Kong has emerged as the world's third largest gold market after

London and New York controlling 10–15 per cent of the total market.

Since before the 1960s, household consumption has shown a tendency towards the pattern of Western developed societies with the amount of income spent on food dropping relative to other expenditures. Although Hong Kong's share of world wealth has increased tremendously over the last three decades, the future of the territory is unclear as much of the wealth was due to the colony's unique relationship with China which shall come to an end in 1997. Whatever happens, Hong Kong will remain a major port for trade between China and the West for many decades after 1997.

## MACAU

Macau was the first port on the China coast to come under the influence of a foreign power and will be the last to return to Chinese sovereignty in 1999 when Portuguese colonial rule ends. Macau is small – less than one-sixty-second the size of neighbouring Hong Kong. The territory is composed of three distinct parts; the Macau peninsula (6.05 sq km), and the islands of Taipa (3.779 sq km) and Coloane (7.087 sq km). Yet the importance of this territory has increased since the mid-1960s due to its rapid economic growth and, since the 1980s, due to its increased role as an open door to China.

### Population growth

Since the 1950s, official statistics of Macau's population have consistently underestimated because of the large proportion of illegal immigrants. In the early 1960s there was a steady flow of Chinese immigrants arriving in the colony and an estimated 75 per cent of these people emigrating again to Hong Kong. In 1989 it was estimated that the population was around 450,300 although a figure closer to 550,000 is probably more accurate.

### Economic change

#### *Urbanization*

In the early 1970s 14 per cent of the population on the Macau peninsula lived in squatter housing (Li *et al.* 1972: 8–16). In recent years, the government has created plans for public housing, tried to control the price of housing by contracts and encouraged the outright purchase of housing by enabling banks to provide credit to buyers.

The greatest change to the development pattern of Macau occurred in 1974 when Taipa was linked to the peninsula by a 2.56-kilometre bridge. This began a period of urbanization on the island which is continuing.

Some ambitious reclamation plans for housing and industry have been announced for areas north-east of the Macau peninsula. However, it is unlikely that these will be started much before 1999.

### Transportation and communications

In 1965 there were ten daily hydrofoil runs between Hong Kong and Macau. By the 1980s vessels were also travelling between Macau and Guangzhou, Jiangmen and Kao-hsiung as well as Hong Kong (Edmonds 1989: xlvii). Now about 6 million people arrive by sea annually with jetfoils carrying over two-thirds of the passengers.

The majority of Macau's shipping is floated over on barges from Hong Kong. A contract to expand Ká Hó Bay on Coloane into a deep water port was signed in 1988.

There is a plan to build a second bridge across to Taipa. In late 1989 reclamation of land and construction of an airport to the east of Taipa and Coloane began which should be completed by 1994 and able to handle jumbo-jets. There have been problems with China over the airport but these now seem to be resolved. A helicopter service between Hong Kong and Macau began in November 1990.

Macau was slow to develop modern telecommunications and the territory's network is currently being modernized. Between 1982 and 1986 the number of telephone lines tripled. Teledifusão de Macau television began broadcasting in May 1984 and there are three radio stations.

### Industrialization and agricultural change

In the early 1960s Macau was an industrial backwater with her most famous industry twelve firecracker factories. Since the early 1970s Macau has experienced rapid growth led by textiles. The reasons for this growth have been cheap labour coupled with *laissez-faire* government. Between 1970 and 1979 textiles experienced a twenty-nine-fold increase in total exports.

Tourism has grown rapidly during the 1980s. Hong Kong residents make up four-fifths of the tourist arrivals. Gambling remains Macau's most popular diversion and accounts for roughly 20 per cent of Macau's gross domestic product. The latest tourist addition will be Macau's first golf club with a 200-room hotel developed with Japanese capital and expected to open on Coloane in 1992.

Since 1983, industrial production has been rising rapidly and diversification away from textiles and casinos began. There has been major growth in artificial flowers, toys, electronic and leather goods. However, textile and garment exports still accounted for just under three-quarters of the value of all exports in 1989. Much of the industrial growth in the late 1980s was due to investors from Hong Kong taking advantage of the cheap unskilled labour, lower land prices, and being able to export goods under a Macau rather than a Hong Kong quota to countries with protectionist tendencies. Shortages of water and electricity dictate that the government will try to promote high technology industries in the future.

Agriculture has virtually disappeared in Macau during the last thirty years with what remains limited to market gardening (Edmonds 1986: 58–60). Today agricultural products, including fresh-water fish, are mostly supplied by China.

## Environmental issues

Development of Macau since the 1970s has put a tremendous amount of pressure on the remaining open spaces. The need to encourage tourism has led to a growing concern for the preservation of historic buildings. Under the direction of the Comissão de Defesa do Património Arquitectónico, Paisajístico e Cultural (Committee for the Preservation of Macau's Heritage) created in 1976, certain areas and buildings have been awarded protected status. Once buildings are given special status, the government gives tax reductions to proprietors as long as they maintain the building's historic character.

As urbanization in Macau has been somewhat less intense than in Hong Kong and the scale of the city is much smaller, warming trends in the territory have been lower (Chiu and So 1986: 103).

The hills on the Macau peninsula are still largely covered with vegetation except for Ilha Verde in the far north-west. On the islands some of the hilly areas are barren although the Macau government is trying to revegetate them. The rapid development of Taipa and Coloane has put the remaining flora and fauna under severe threat.

Rapid development without any waste water treatment facilities has lead to increasing water pollution. A water treatment plant and sewer construction are now underway. In May 1990 the government had to impose fines on anyone fishing for any species or collecting shellfish or seaweed in most of the waters adjacent to the territory (*Tribuna de Macau*, 2 June 1990: 19). One waste dump in the north of the peninsula has already been filled and converted into a park but there are worries of toxic wastes polluting the surrounding waters ('Aòmén shēngtài' 1990: 4).

A citizen's action group, a government standing committee on environmental problems and an environmental planning group have been formed and a draft environmental law has been recently passed.

## Cultural context

The people of Macau fall into three major groups: the Macaense, the Cantonese and the Portuguese. Most people consider anyone of mixed ancestry who speaks Portuguese as a Macaense. The language of the Macaense is a Portuguese creole which along with the cuisine and customs of this people is rapidly fading.

The Cantonese make up roughly 85 per cent of the territory's population. Although linguistically and in some ways culturally distinct, the Cantonese view themselves as part of a larger Chinese ethnic group composed of Fujianese, Shanghainese and other Chinese found in the territory (Cremer 1991: 126–7).

The number of Portuguese in Macau has remained around 2–3 per cent since the 1920s. It is doubtful whether Portuguese culture has much of a future in Macau. The inclusion of Portuguese as an official language in Macau during the fifty-year 'transition period' after 1999 is one achievement.

In addition to various Chinese beliefs, there are approximately 30,000 Catholics. In relation to Macau's population growth, however, expansion of the Catholic congregation has been small.

## Political change

With the recognition by China of Portuguese sovereignty over Macau in 1887, the legal system began to undergo a process of amalgamation which continued until the 1970s. The anti-government riots in Macau during 1966 and 1967 along with the *coup d'état* in Portugal during 1974 all helped to begin the reversal of this integration process. The promulgation of the *Estatuto Orgânico de Macau* (Macau Organic Statute) in 1976 can be taken as the turning point in Macau's constitutional make-up as this statute gave the territory great political autonomy. The Sinification of Macau's political system began in 1979 when China and Portugal concluded a secret agreement defining Macau as Chinese territory under Portuguese administration. Sinification has slowly intensified since the start of negotiations in 1986. Cantonese language will have complete equal status with Portuguese by 1999. Macau's air and sea borders with China have never been clearly defined with a line midway down the Porto Interior and the channel between Taipa and Coloane and the Chinese islands of Xiaohengqiin and Dahengqin serving as the unofficial western boundary. The Sino-Portuguese joint Liaison Group was discussing finalization of the boundary question as of mid-1991.

The Macau *Assembleia Legislativa* is controlled by the Portuguese government and the Chinese business and civic organizations through appointments of the majority of deputies. The law-making process itself is extremely complex, with both Portuguese and Macau bodies possessing legislative power.

## International context

Since 1970 the market for Macau's exports has changed significantly. Exports to the USA rose during the second half of the 1970s. By 1989 the USA, Hong Kong, West Germany, France and the UK remained important trading partners while Portugal, Africa, Latin America and the Soviet bloc were insignificant markets.

Imports can be difficult to measure as much passes through Hong Kong. Hong Kong and the People's Republic of China together accounted for more than 60 per cent of Macau's imports in 1989. Japan, the USA and the EC are other major sources of imports.

Since the violent anti-government demonstrations by pro-Communist Chinese residents during 1966–7, Portuguese authority weakened and Beijing has had a far stronger say in Macau. After the 1974 *coup d'état* in Portugal, the Portuguese government apparently informed the People's Republic of China of its intention to leave Macau. China refused to discuss the issue. In March 1987 Macau's fate was decided by a joint Sino-Portuguese declaration. The current status will be maintained until 20 December 1999 when administration of Macau will be taken over by the People's Republic of China. However, Macau will maintain some independent status in international organizations. The territory joined the General Agreement on Tariffs and Trade (GATT) as its 101st member in 1991 and recently joined the International Maritime Organization.

## QUESTIONS FOR FURTHER RESEARCH

The key to Taiwan's future lies in its relation to mainland China. Within such a large topic, one area worthy of further research is the increasing impact of Taiwan upon the mainland as a result of the loosening of travel restrictions in the latter half of the 1980s. In the international sphere there still remains the problem of the Tiao-yü-t'ai (Diaoyutai, in Japanese, Senkaku) islands to the north-east of Taiwan which are claimed by China (both Beijing and T'ai-pei) and Japan and requires further research into territorial claims.

The role of Hong Kong and Macau in introducing European (primarily British and Portuguese) and Taiwan in introducing North American ideas into the post-1978 'open' mainland is critical and worthy of more research than it has received in recent years. Hong Kong's role has been looked at primarily from an economic viewpoint, whereas the Portuguese have written a considerable amount on Macau's cultural influence on China – particularly from a historical perspective. One topic of continuing interest to geographers should be the relationship of these two territories to their Guangdong hinterlands. Likewise, Taiwan is beginning to develop a similar 'special relationship' with Fujian Province across the straits.

All three territories are serving as models for development of Hainan Island. The mainland Chinese have said that they wish Hainan would become a 'second Taiwan', Hong Kong investment is taking the lead in Hainan, and casino operations have been discussed which would be developed on a Macau model. The possibilities for research in this area are tremendous.

Taiwan's internal political and environmental problems also pose interesting questions for the future. Will Taiwan evolve towards independence, will she be integrated into the People's Republic, or will social instability on the mainland give the island a chance to play a major role in the formation of a new China?

There is considerable discussion as to appropriateness of the geographical divisions within Taiwan Island. Research needs to be done on the feasibility of the proposed combining T'ai-chung City and T'ai-chung County to form a third directly administered city. Can Taiwan solve the pressing environmental problems which the island faces through the application of high-tech solutions which the government's surplus should allow, will these funds be used for other purposes, or do the population densities on the island necessitate a low-tech, think-small solution?

The major areas for research on Hong Kong and Macau proper are related to the impact of Chinese rule on the territories after the late 1990s. For geographers the most important predictive areas revolve around the impact of changes in local policy and the relationships of the territories to their Guangdong hinterlands after reversion. The assessment of the United Kingdom and Portugal's roles after the late 1990s is also a topic worthy of investigation. In particular it will be interesting to see what changes in land use occur inside Hong Kong and Macau due to the new political order.

Whatever the future holds, there is little doubt that these territories will play an increasing role in the future of China. The key question now is whether this is a role these territories wish to play.

## REFERENCES

'Aòmén shēngtài zāoshòu pòhuài jídài fǎ "zhì" héshān' (1990) Zhōngguó huánjìng bào 880 (25 December): 4.

'Běi Shì rénkǒu mìdù yuèjū shìjiè dìyī' (1991) Zhōngyāng rìbào (International Edition) 22808 (3 April): 7.

Bristow, R. (1989) Hong Kong's New Towns: A Selective Review, Hong Kong and Oxford: Oxford University Press.

Chén Yìyí (ed.) (1989) Zhōnghuá mínguó huánbǎo fǎguī, T'ai-pei: Jīnyù Chūbǎnshè.

Chiu, Tse Nang and So, C.L. (eds) (1986) A Geography of Hong Kong, 2nd edn, Hong Kong and Oxford: Oxford University Press.

Council of Agriculture (1986a) Agricultural Development in the Republic of China Taiwan – A Graphic Presentation, T'ai-pei: Council of Agriculture.

Council of Agriculture (1986b) First General Report of the Council of Agriculture, T'ai-pei: Council of Agriculture.

Council for Economic Planning and Development (ed.) (1989) *Taiwan Statistical Data Book 1989*, T'ai-pei: CEPD.

Cremer, R.D. (ed.) (1991) *Macau: City of Commerce and Culture*, Hong Kong: API Press.

Edmonds, R.L. (1986) 'Land use in Macau: changes between 1972 and 1983', *Land Use Policy* 3(1): 47–63.

Edmonds, R.L. (1989) *Macau*, Oxford: Clio Press.

Government Information Office (1987) *Republic of China 1987 Reference Book*, T'ai-pei: Hilit.

Government Information Services (1989) *Hong Kong 1989*, Hong Kong: GIS.

'Hángtài děng 10 dà xīnxìng gōngyè jí bāxiàng guānjiàn jìshù nàrù guójiā jiànshè 6 nián jìhuà' (1990) *Zhōngyāng rìbào (International Edition)* 22696 (7 December): 2.

*Hong Kong Monthly Digest of Statistics* (monthly) Hong Kong: Census and Statistics Department.

'Huánbǎofángwū jiāngchéng 90 nián míngxīng qìyè' (1990) *Zhōngyāng rìbào (International Edition)* 22665 (6 November): 7.

Huáng Bìxiá (1991) 'Míngdé Shuǐkù shǐyòng shòumìng liàng hóngdēng' *Zhōngyāng rìbào (International Edition)* 22761 (10 February): 7.

Huáng Chūnmíng (1990) 'Chéngxiāng bǐjì', *Zhōngyāng rìbào (International Edition)* 22699 (10 December): 4.

Li Cheng and White, L. (1990) 'Elite transformation and modern change in mainland China and Taiwan: empirical data and the theory of technocracy', *China Quarterly* 121: 1–35.

Li, Kwoh Ting (1988) *The Evolution Policy behind Taiwan's Development Success*, introduction by Ranis, G. and Fei, J.H.C., New Haven, Conn., and London: Yale University Press.

Li, Shu-fan, Poon, Sheung-tak, and Yeung, Chee-on (1972) 'Housing in Macau', *Annals of the Geographical, Geological and Archeological Society, Hong Kong University* 1: 8–16.

Lorot, P. and Schwab, T. (1986) *Singapour, Taiwan, Hong Kong, Corée de Sud, Les Nouveaux Conquérants?*, Paris: Hatier.

'Nóngyè shēngchǎn yīngyǐ nèixiāo wèi mùbiāo' (1990) *Zhōngyāng rìbào (International Edition)* 22699 (10 December): 7.

Qín Kěxǐ (ed.) (1989) *Běnguó dìlǐ kèwén biǎojié*, T'ai-pei: Zhōngguó Qīngshǎonián Chūbǎnshè.

'Quánguó huàfēn wèi 10 ge shēnghuóchuàn' (1990) *Zhōngyāng rìbào (International Edition)* 22717 (28 December): 2.

Rèn Wéixìn (ed.) (1990) *Běnguó dìlǐ tújí jīngxī*, T'ai-pei: Dōngshān Chūbǎnshè.

Research, Development and Evaluation Commission (1988) *Annual Review of Government Administration, Republic of China*, T'ai-pei: The Executive Yuan.

Serviço de Meterorologia e Geofísica (1985) *Tempo em Macau: Weather in Macao: Aòmén Tiānqì*, Macau: SMG.

'Shǒuzuò xiūxián nóngyè nóngchǎng' (1991) *Zhōngyāng rìbào (International Edition)* 22812 (7 April): 7.

'Táiběi Shì wèi quánqiú kōngqì wūrǎn zuìqīng chéngshì?' (1991) *Zhōngyāng rìbào (International Edition)* 22809 (4 April): 7.

The Steering Committee Taiwan 2000 Study (1989) *Taiwan 2000: Balancing Economic Growth and Environmental Protection*, T'ai-pei: Institute of Ethnology, Academia Sinica.

*Tribuna de Macau* (1990) Ano 8 (398), 2 June, Macau: Tribuna de Macau.

Williams, J. (1988) 'Vulnerability and change in Taiwan's agriculture', *Pacific Viewpoint* 29(1): 25–44.

Winckler, E.A. and Greenhalgh, S. (eds) (1988) *Contending Approaches to the Political Economy of Taiwan*, Armonk, NY and London: M.E. Sharpe.

Wu Qi (1988) *South Korea and Taiwan: A Comparative Analysis of Economic Development*, Brighton: Institute of Development Studies, University of Sussex.

# 7

# THE CHANGING GEOGRAPHY OF JAPAN

*Edwina Palmer*

## INTRODUCTION

The 'economic miracle' of the speed of Japan's recovery from the devastation of the Second World War received much attention world-wide from the 1960s. Economic growth in Japan since then has interacted with changes in the population, technology, the natural environment and Japan's relations with the rest of the world.

## OLD PEOPLE, CITY PEOPLE

Arguably, people are Japan's most abundant natural resource and it is the changes in the structure and distribution of the population which has most fundamentally altered the geography of Japan since the 1960s (see Figure 7.1). The increase in population, changes in vital statistics, the population pyramid and the number of households from 1960 are shown in Figures 7.2, 7.3 and 7.4. Concomitant with these changes was a marked increase in the expectation of life at birth, from 63.6 years for males and 67.8 for females in 1960 to 75.5 for males and 81.3 for females in 1988; by 1987, the Japanese were overall the longest-lived people on earth. The proportion of the aged (over 65 years) to the total population reached 7 per cent around 1970, qualifying Japan as an 'aged society' by UN criteria. In fact, Japan's population is ageing more rapidly than that of any other nation, and by 1990 some 12 per cent were aged over 65 years. The problem of how to cope with becoming a 'super-aged' society, in which the over-65s are not uncommonly caring for over-85s, is a matter of considerable national concern.

Accompanying the rapid ageing of the population has been a

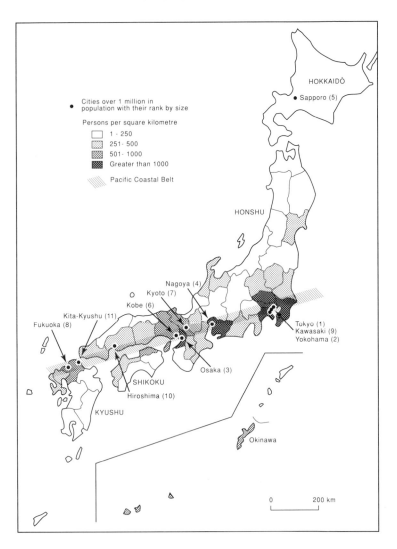

*Figure 7.1* Administrative regions and population density in Japan 1988

trend away from the traditional stem family known as the *ie*, in which the elderly were cared for usually by their eldest son and his wife. Average family size has shrunk as young couples increasingly prefer privacy and independence from parental influence, as parents prefer to limit family size to only two or three children, and as

196

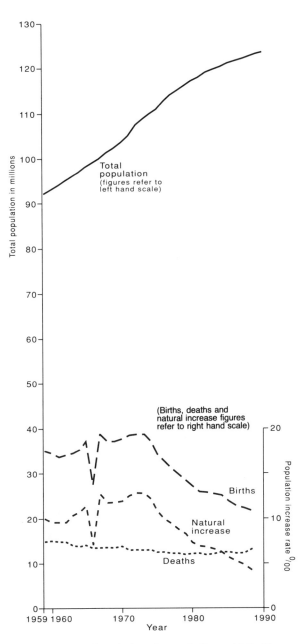

*Figure 7.2* Population change in Japan 1960–89

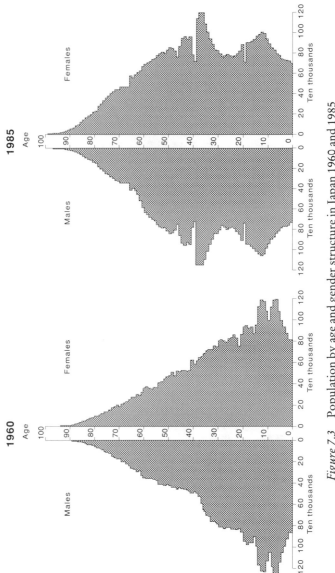

*Figure 7.3*  Population by age and gender structure in Japan 1960 and 1985

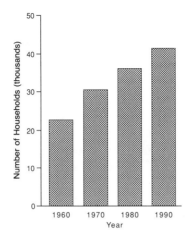

*Figure 7.4* Changes in the number of households in Japan 1960–90

urban housing units are not designed for large families. While in 1960 there were 22.6 million households (see Figure 7.4) averaging 4.5 persons each, by 1990 the number had doubled to 41.0 million households, averaging only 3.0 persons each. The typical family is now therefore a nuclear family with dependent children. The rapid rise in the number and proportion of old people living alone is a new social phenomenon, and adaptation to the special needs of the elderly in the future is steadily effecting changes in the pension and welfare systems and in the medical and leisure sectors. It is anticipated that in addition to the direct costs on the economy in the future, in terms of higher pension and medical bills, the economy is also likely to be affected by ageing in less obvious ways. The seniority system (*nenko joretsu*) of higher pay for longer service prevalent in most sectors of economic activity meant that rapid economic growth in the 1960s was buoyed up by relatively cheap labour from a youthful labour force. Unless there is a fundamental change in this custom, costs for labour are likely to rise markedly as the labour force ages. There was some evidence during the economic boom of the late 1980s that competition among employers for new recruits may indeed bring about such a change in due course.

In 1950 Japan's population of some 83.2 million was widely dispersed throughout the country, with almost half (45.4 per cent) living in farming households in rural areas. During the period of high economic growth rates from the late 1950s and especially

1964–73, urbanization occurred very rapidly as manufacturing industries expanded, mainly along the so-called Pacific Coastal Belt from Tokyo to Northern Kyushu (see Figure 7.1). Almost all school-leavers in rural areas close to the Pacific Coastal Belt were rapidly drawn to jobs in manufacturing and service industries from the mid-1950s. This outflow steadily spread outwards to more remote areas, to peak in Hokkaido around 1970. From 1970 to 1990 some 46.8 per cent of the total land area was designated as 'depopulated' and eligible for special funding; by 1985 it contained only 6.7 per cent of the total population, with an average density of 47 per sq km, of whom 17 per cent were aged over 65 years.

Whereas there were only six cities with a population of more than 1 million in 1960, the figure had almost doubled to eleven cities by 1988, and all apart from Sapporo were along the Pacific Coastal Belt (see Figure 7.1). From 1970 the National Census included data on Densely Inhabited Districts (DIDs), revealing that 53.5 per cent of the population were already concentrated on only 1.7 per cent of the land area, (excluding Okinawa), especially around Tokyo, Osaka and Nagoya.[1]

Problems of congestion and pollution, combined with the economic effects of the Oil Crises of 1973 and 1978, encouraged outmigration from these city centres throughout the 1970s and contributed to both the marked growth of provincial capitals such as Sapporo and the outward expansion of existing cores along the Pacific Coastal Belt. Despite government attempts to assist deconcentration of population and industry, centralization continued with renewed vigour in the 1980s, so that the Pacific Coastal Belt became in effect one continuously built-up area. Most conspicuous was accelerated concentration into the Kanto region centred on Tokyo especially from 1986, resulting in spiralling land values. The average population of DIDs in Japan was nearly 7,000 per sq km in 1985, but for those in Tokyo Prefecture it was over 11,500 per sq km. By 1989 42.8 per cent of Japan's total population resided within a 50-kilometre radius of Tokyo, Osaka and Nagoya, and little less than a quarter (23.2 per cent or 28.6 million) within the same distance from central Tokyo alone.

The increasing demand for land in these regions was but partially met by large-scale coastal reclamation projects, especially in Tokyo and Osaka Bays.

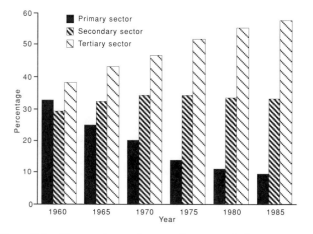

*Figure 7.5* Changes in employment by sector in Japan 1960–85

## ECONOMIC RESTRUCTURING

Figure 7.5 indicates that structural change in the Japanese economy proceeded rapidly from 1960, with a decline in the importance of primary industries to a level approaching that of the West, and a rise to prominence of the tertiary sector. What this does not reveal is that while there was relatively little restructuring of agriculture, major structural adjustments were made in manufacturing and service industries.

### Agriculture: from rice bowl to gourmet cuisine

With increasing affluence, Japanese consumer demand in virtually all areas of consumption including foodstuffs developed through three stages: 'quantity' in the 1960s, 'quality' in the 1970s and 'luxury' in the 1980s. In order to ensure adequate and equitable supplies of rice on the one hand and to guarantee farm incomes on the other, after the Second World War the government compulsorily purchased all the rice harvest and redistributed it to consumers at below cost price. As income from rice cultivation was thus secure, and as technological advances took place such as the invention of labour-saving machinery in the 1960s, rice cultivation became by far the most stable and profitable sector of farming. Both production and productivity increased markedly during the 1960s. At the same

201

time, the increasingly urban, affluent and Westernized consumers were purchasing a greater diversity of foodstuffs, and demand for rice fell from 314.9 gr to 194.4 gr per capita per day between 1960 and 1988. For political reasons, it was expedient to continue with the rice price support policy, and the result was a massive over-supply of rice that was too expensive to export. The cost to the government in 1974 was Y600,000 million. The stockpile peaked in 1980 but declined thereafter and the government finally broke even in 1987. This was achieved by steady reduction of the acreage under rice, by some 15 per cent 1970–88, mainly through diversification and fallowing, and by lowering the producer price from 1986. Nevertheless that was still six times more than the price to American producers and thirteen times more than in Thailand. It is not only rice-exporting nations that are demanding liberalization of the Japanese rice market but also the Japanese food processing industry.

Rural depopulation almost halved the farming population between 1960 and 1989, and reduced the number of farm households by one-third. However, continued depopulation has affected farm size very little, since former farm families typically retain ownership of their land as an investment. Farm size remains a major obstacle to the viability of agriculture, with some 95 per cent of farms being less than 3.0 ha (1989). Thus Japanese farms have been kept artificially viable by strong protectionist policies. The 1961 Agricultural Basic Act aimed to increase farm incomes through 'selective expansion' (particularly of beef, fruit and vegetables) and greater participation in co-operatives. Farm incomes did indeed rise, but more through participation in off-farm jobs. Whereas in 1960 approximately one-third (34.3 per cent) of farm households were full-time farmers, by 1970 the proportion had fallen to 15.6 per cent and it stabilized at around 12.5–14.5 per cent thereafter.

As consumers became more affluent, tastes in diet became more diverse and overall self-sufficiency fell from 91 per cent in 1960 to 70 per cent in 1988. In particular, Japan was only around 6 per cent self-sufficient in soy beans and 16 per cent in grains other than rice (1988). From 1981, with the trend to luxury in consumption in general and fashion for 'health food', per capita consumption of vegetables exceeded that of cereals, to stabilize at around 300 gr daily. Nevertheless, Japan remains largely self-sufficient in vegetables, partly due to increased covered cultivation in 'vinyl houses'.

Under pressure from the United States to reduce the imbalance of trade, liberalization of imports of beef and fresh oranges from 1991 and orange juice from 1992 has been agreed. Unable to compete, some restructuring and further retraction of the farming sector is to be expected thereafter.

## Industry: from ships to microchips

Japan is notably lacking in natural resources for large-scale manufacturing industries and dependence upon imported raw materials is therefore very high (see Figure 7.6). The 'economic miracle' of the 1960s was founded on the importation of raw materials, and manufacturing and export of finished products. Heavy industry flourished, mainly on coastal locations, including purpose-built sites reclaimed from the sea, close to urban labour and consumer markets and deep-water ports. Importation of coal, crude oil, metal ores, limestone and salt supplied steel works, chemical and petro-chemical industries that were often integrated on one site. These in turn supplied the ship-building, automobile, textile and other light manufacturing industries with an eye to exports. Largely thanks to the profitability of these industries, real economic growth rates of more than 10 per cent were sustained throughout the 1960s; and the Tokyo Olympic Games of 1964 are often cited as stimulating the economy further by an accompanying boom in the construction industry.

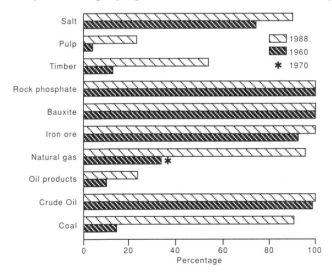

*Figure 7.6* Degree of dependence on imported minerals and fuels in Japan 1960 and 1988

203

The profitability of this economy was due primarily to the cheapness of imported raw materials, especially crude oil, the availability of a relatively young, cheap and skilled labour force, and healthy economies overseas eager to accept goods exported from Japan. All three of these conditions altered rapidly in the 1970s. The rapidity with which industrialization occurred during the late 1950s and early 1960s soaked up surplus labour and produced a labour shortage by the late 1960s, causing an increase in labour costs. The Oil Crises of 1973 and 1978 had even more far-reaching effects, by raising the cost of imports, by causing a world slump in shipping that arrested Japan's ship-building industry, and by causing retraction of the consumer markets in overseas destinations for exports.

To cope with this situation, restructuring of manufacturing industries took place, with a marked trend away from heavy industries to light industries such as cameras and household appliances (see Figure 7.7). The so-called 'technology gap' between Japan and the West had already been largely closed by 1965. After the first Oil Crisis, international competitiveness was maintained in traditionally labour-intensive sectors such as the automobile industry by the increasing application of high technology such as robotics. In particular, with the availability of a highly educated workforce, knowledge-intensive, technologically advanced sectors of manufacturing expanded rapidly to produce semi-conductors, integrated circuits, computer hardware, optical and medical equipment, machining centres (computer-controlled lathes) and 'new materials' such as fine ceramics. The value of production of integrated circuits grew by more than 30 per cent per annum from 1975 to 1985.

Over-capacity in heavy industries was dealt with from around 1983 by, for example, the closure of ore mines, and the disposal of surplus plant and equipment. This latter occurred in sectors such as steel, petrochemicals and cement; some such as the steel industry went on to recover profitability in the late 1980s.

However, the economy was still primarily geared towards export of finished products. To alleviate trade friction, Japan agreed virtually to double the value of the yen against the US dollar in 1985. The aim was to curb exports, promote imports and stimulate economic growth primarily through domestic demand and consumption. The success of this strategy exceeded expectations and the result was the longest post-war boom in the business cycle

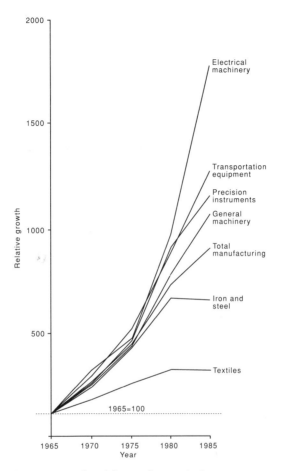

*Figure 7.7* Industrial growth rates in Japan 1965–85

except for the period 1965–70. Real economic growth, which had been maintained at 3–5 per cent since 1973, was sustained even after revaluation of the yen in 1985.

Public investment in social overhead capital and availability of low interest loans stimulated the construction industry, and spiralling land values, especially in central Tokyo, prompted large-scale rebuilding. Consumer demand shifted further from 'quality' to 'luxury', and sales soared of expensive prestige consumer durable items such as large televisions, silent washing machines and European cars. Manufacturers for the domestic market therefore moved further into production of high quality, highly processed, high value-added goods, especially electronics.

On the other hand, manufacturers in more traditional sectors rapidly sought to retain international competitiveness in the face of the rise of the value of the yen, trade friction and competition from newly industrializing economies (NIEs) in Asia, especially Korea, by a flurry of amalgamations, joint ventures and foreign direct investment. Examples of the latter include the buying up of operational foreign non-ferrous metal ore mines; the construction of a new steel works in Shandong Province in China by five main Japanese steel manufacturers to produce steel plates for the automobile industry; and the establishment of no fewer than fourteen automobile assembly plants overseas between 1983 and 1989, mostly in North America.

## The tertiary sector: at your service

The tertiary sector has been the fastest-growing sector of the economy, especially since the mid-1980s. The huge trade surplus brought about an accumulation of foreign currency in Tokyo which raised the prominence of Tokyo as an international financial market during the 1980s. Being fortuitously located mid-way between the New York and London financial markets, this contributed to the globalization of round-the-clock financial dealing. Tokyo became the world's second largest financial market after New York in 1986. Activity is concentrated in Chiyoda ward, giving rise to the 'City Phenomenon' named after the concentration of financial activity in the City of London. The increase in the importance of Tokyo as a leading financial market contributed greatly to the further concentration of population, capital and central management functions into the Tokyo region during the 1980s, and the accompanying increase in land values.

However, while this aspect of Japan's tertiary sector received much attention overseas, it was in fact personal services which grew most rapidly, particularly business services such as consultancies.

## Imported energy: the constant anxiety

As in the case of other raw materials, Japan is resource-poor in carboniferous deposits suitable for the generation of electricity. At the beginning of the period of rapid economic growth, in 1955, one half of Japan's energy requirements were met by coal and one-fifth each by hydro-electricity and oil. Domestic deposits, mainly of coal in Hokkaido and Northern Kyushu, were viable, so that Japan was dependent on imports to supply only one-quarter of required energy.

However, with rapid industrialization, demand for energy more than quadrupled from 930 billion kcals in 1960 to 4,455 billion kcals in 1988. Consumption of oil rose by as much as 21.4 per cent per annum throughout the 1960s, so that by 1970 Japan was 70.8 per cent reliant on oil for energy, all of which was imported. The Oil Crisis of 1973 brought about perception of the need to develop domestic energy supplies and to diversify both the types of raw materials for energy generation and the sources of imports of them. Dependence on oil was thereby reduced to 57.3 per cent in 1988; approximately two-thirds was obtained from the Middle East.

Coal continued to supply one-fifth of energy requirements from 1970, but Japanese coal mines became increasingly unviable; whereas 45 per cent of coal was domestically produced in 1970, the proportion fell to 11 per cent by 1988. The alternative sources of energy selected have been natural gas – mostly imported – and nuclear power generation, each of which supplied about 10 per cent of requirements during the late 1980s. Nuclear power is particularly regarded as a consistent potential supplier of energy in the future. By 1989 there were thirty-seven reactors in operation, thirteen under construction and three more being planned. However, the earliest reactors are expected to reach the end of their operational lives in the late 1990s.

Demand for energy appears to be still growing, and Japan was around 90 per cent dependent upon imports for energy supply throughout the 1980s. The potential of solar energy and burning of household refuse for thermal power generation are two of the alternative sources being researched in the attempt to be more self-sufficient in energy.

## Transports of delight?

Upgrading of the transportation system has been a perennial priority in Japan since the mid-1950s, at first to facilitate the movement of goods and people in order to increase economic efficiency along the Pacific Coastal Belt. By the mid-1960s, development of the urban transportation network was urgent as a countermeasure to traffic congestion. And from the 1970s there was the further aim of developing transportation in rural areas in order to stimulate regional economies and help counteract rural depopulation.

The Shinkansen ('Bullet Train Line') has been the showpiece of Japanese transportation technology, and its development largely mirrors the above-mentioned priorities. It was first opened in 1964 to coincide with the Tokyo Olympic Games, between Tokyo and Osaka. The distance of 515 kilometres was to be covered in less than three hours, making it the fastest train ride in the world, a record that was not to be surpassed for nearly two decades until the opening of the French TGV in 1981. The success of the Shinkansen was such that the Diet voted in 1970 to form a nationwide network of Shinkansen lines, and work proceeded to extend west of Osaka to Okayama in 1972 and Hakata in 1975. The northern extensions were delayed, partially due to the effects of the Oil Crises. Sections from Omiya to Morioka and Niigata were opened in 1982, and from Ueno (central Tokyo) to Omiya in 1985. By the late 1980s the expansion of the Pacific Coastal Belt had turned the Tokyo-Osaka stretch of the Shinkansen from a luxury excursion route into a daily commuter belt, with trains carrying 1,440 passengers at 220 kph, running to capacity at approximately ten-minute intervals (115 return journeys per day). Demand having outstripped safe supply, the construction of a 'Central Linear Shinkansen' is undergoing trials in Yamanashi Prefecture 1990–4, with the aim of linking Tokyo and Osaka within one to two hours by linear motor car before the end of the twentieth century.

By contrast with the speed, comfort and excellent safety record of the railways, the provision and standard of roads is conspicuously lagging. The rate of ownership of four-wheeled motor vehicles, at 43 per 100 people, was similar to that of Britain or Italy in 1989. However, only 22 per cent of the 1.1 million km of roads were wide enough for two buses to pass. Urban traffic congestion continues to be a major problem. While motorway construction continued, it failed to keep pace with the rapid increase in private vehicle ownership and urbanization. The urgency of the need for improvement in the motorway network has led to plans to double the total length from 4,280 km (1988) to 9,000 km during the 1990s. As manufacturing tended to move away from requiring shipment of bulk raw materials by sea to high value-added finished products, so it increasingly relied on trucks for haulage. They accounted for 51 per cent of freight handled in 1988. Moreover, with the trend to luxury and personalized services in the 1980s, courier and door-to-door deliveries by light van became popular, adding to the congestion of the roads.

The late 1980s saw the completion of several major transportation projects. With the completion of the Hokuriku Expressway in 1988, all regions of Honshu were connected by a continuous motorway network. The Seikan Tunnel, the world's longest undersea tunnel, some 53.9 km long, opened to trains in 1988 to link Honshu and Hokkaido. Also in 1988, the 13.1 km Seto Ohashi Bridge was opened, creating a direct road and rail link between Shikoku and Honshu. Thus all four main islands are now connected by road and/or rail.

At the regional scale, several projects are intended to relieve the appalling traffic congestion around Tokyo Bay. Yokohama Bay Bridge, 2.8 km long, was opened in 1989 and markedly relieved congestion in Yokohama city. Construction of a Trans-Tokyo Bay Highway commenced in 1989, with an anticipated completion date of 1996. Part tunnel and part bridge, this 15.1 km expressway will link Kawasaki to Kisarazu and will ultimately form part of a ring road round Tokyo. While it is expected to relieve congestion in Tokyo, there are serious concerns over the effects of its construction upon the natural environment of Tokyo Bay.

The air transportation network has been no more able to cope with increased traffic than have the roads. As of 1990 there were only three major international airports: two in Tokyo and one in Osaka. Congestion at Tokyo International Airport (Haneda) resulted in the planning of a new airport at Narita from 1962. Completed in 1973, protestors prevented its opening until 1978. Despite having only one runway, New Tokyo International Airport (Narita) handled the most freight and the eighth largest number of passengers of any airport in the world in 1988. The marked increase in the volume of freight during the 1980s is attributed to the increased prominence of high value-added products and shipment of fresh foodstuffs, especially seafood and vegetables, throughout the 1980s.

Congestion remains a problem, and new runways are under construction: in the case of Haneda Airport on land being reclaimed from Tokyo Bay. In Osaka Bay, work is in progress for the building of the New Kansai International Airport, the first anywhere in the world to be built in the sea, to minimize nuisance from aircraft noise. The site was at an original sea depth of 20 metres. Already facing increased costs due to unanticipated problems of subsidence, it is unlikely to be opened before 1994, and the planned three runways have been reduced to an initial one.

Provincial airports have been subject to no less controversy. The New Amami Airport in Kagoshima Prefecture, for example, was opened in 1988. While on the one hand it was expected to boost local tourism, its construction involved the destruction of a coral reef. The dilemma of the conflict between the practicalities of service provision and the preservation of the precious is nowhere so acutely evident as in Japan's transportation problems.

## MINAMATA AND AFTER: ENVIRONMENTAL ISSUES

As a result of largely unrestrained urbanization and expansion of manufacturing industries during the 1950s and 1960s, environmental pollution inevitably resulted in a variety of ways – some particularly tragic – which drew the attention of the world's press. The most notorious is the outbreak of 'Minamata disease' in Kumamoto Prefecture from 1959, in which industrial effluent containing organic mercury poisoned fish stocks in Minamata Bay that in turn caused irreversible damage to the nervous systems of those who ate the fish. A similar case of mercury poisoning occurred in Niigata Prefecture in 1965. As a result, nearly 3,000 victims of Minamata disease were officially recognized. More than 1,000 of these had died by 1990, and the courts were still dealing with claims for recognition and compensation from more than 3,000 others.

From 1955 residents of the Jinzu Valley in Toyama Prefecture presented symptoms of kidney damage and bones so fragile that even coughing caused fractures, a condition which became known as *itai-itai* ('It hurts, it hurts') disease. The cause was traced in 1964 to cadmium poisoning from effluent from a local metal ore mine. Of 128 officially recognized victims, 113 had died by 1989.

While the effects of environmental pollution upon individuals are clearly sometimes fatal or irreversible, they are sometimes no less so upon the environment itself. Continuous operations at Ashio Copper Mine in Tochigi Prefecture from 1884 caused acid rain locally from sulphur dioxide and completely destroyed all flora of what was once surrounding forestland. Refining ceased in 1956 and mining was halted in 1973, since when there has been a serious attempt at reafforestation of the area. Progress is slow, despite an annual budget in the late 1980s amounting to Y300 million per annum.

The shocking incidents of mercury and cadmium poisoning, as well as recognition of other pollution related conditions such as Yokkaichi asthma, raised public awareness of environmental issues, which prompted spates of 'citizens' movements' in the late 1960s. Issues tackled were various, including not only those mentioned above but also noise and vibration from newly constructed express-ways and Shinkansen lines, aircraft noise near airports, and un-pleasant odours from farms, factories and service industries. Atmospheric pollution from motor vehicle exhaust fumes became of such concern that some traffic officers on point duty in central Tokyo were issued with oxygen masks.

Activity by citizens' movements largely abated during the early 1970s after the introduction or amendment of various laws pertain-ing to industrial and environmental standards, and there were no major incidents of the Minamata kind during the following two decades. However, as attempts were made to decentralize manu-facturing industries from the 1970s on, light industries were encour-aged to locate in rural areas. High-tech industries such as electronics were welcomed for their 'clean' non-polluting image. While no cases have yet been proven, there was concern in the 1980s that the large quantities of trichloroethylene used in washing integrated circuit boards may be carcinogenic and harmful to the environment.

Close monitoring revealed that the biological oxygen demand of streams in urban areas fell steadily from 15.23 ppm in 1975 to 7.12 ppm in 1987, showing a demonstrable improvement. Likewise, the incidence of 'red tides' (blooms of toxic algae) in the Inland Sea decreased from 255 in 1975 to 107 in 1987. But while the concentra-tion of sulphur dioxide in the atmosphere at the fifteen monitoring stations fell markedly between 1965 and 1988, the amount of nitrogen dioxide actually increased to 0.042 ppm in 1988, one of the worst levels on record. This is attributed to increased motor vehicle exhaust associated with the boom in automobile sales in the late 1980s.

Urban growth and industrial development necessitated increased extraction of ground water, and subsidence not infrequently occurred as a consequence in major cities. Parts of Tokyo subsided cumula-tively more than four metres over the past century, and subsidence was a notable problem in Tokyo and Osaka especially from 1950 to 1965. With the outward expansion of the Tokyo agglomeration, subsidence likewise affected a widening area, mainly the northern Kanto Plain from 1973 to 1985.

In fact, several environmental issues surfaced in Japan in the late 1980s; but whereas other nations expressed concern over Japanese economic activities that affected the global environment such as whaling, driftnet fishing and 'pollution export' with the establishment of factories overseas, in Japan itself the problem of the disposal of waste, both industrial and domestic, was of more urgent concern. A corollary of affluence and 'luxury' consumer consciousness in the 1980s was excessive packaging of virtually all consumable items, especially foodstuffs. This trend, combined with the renewed concentration of the population into major centres, caused increasing concern among local authorities responsible for refuse disposal, especially from 1985. The total volume of refuse produced

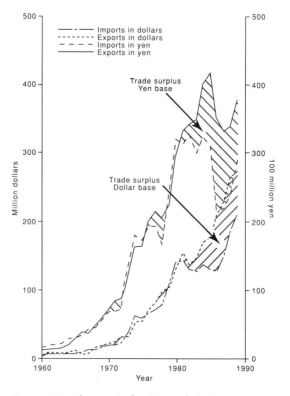

*Figure 7.8* Changes in foreign trade in Japan 1960–90
*Source*: Yano Tsuneta Kinenkai (1990: 589)

212

by Tokyo City (twenty-three wards) rose by a quarter from 3.8 million tonnes in 1983 to 4.8 million tonnes in 1988. City incinerators could no longer cope with all the burnable rubbish, of which some 20 per cent was dumped unburnt along with non-burnable waste on coastal reclamation landfill sites. A plague of flies in the area in 1989 was attributed to this. Dumping sites are all expected to be filled by around 1993, and authorities are at a loss as to how to safely dispose of refuse thereafter.

Safe disposal of industrial waste, in particular, is of considerable concern. Already Tokyo exports industrial waste to rural disposal centres as far away as Nagano Prefecture, Shikoku Island and Hokkaido. They, too, are rapidly reaching capacity, and local residents are likely to resist any attempts to establish new dumps. Cases of illegal dumping of industrial and other waste are expected to increase.

Thus the question of recycling has of necessity become a topic of debate: not so much through concern over resource depletion as through the problem of waste disposal. Of special concern is the recovery of the mercury and cadmium contained in the 2,500 million or so dry cell batteries consumed in Japan annually, disposal of which until now is thought to be causing low-level and diffuse environmental contamination: these very substances were the causes of Minamata and *itai-itai* diseases.

## TRADE AND BALANCE OF PAYMENTS: AN EMBARRASSMENT OF RICHES

Japan has enjoyed a trade surplus since 1964 (see Figure 7.8). Along with structural changes occurring in the Japanese economy since then, there were concomitant shifts in the pattern of foreign trade. Most notably, there was a trend away from dependence on importation and processing of raw materials for export as finished products. Although in 1989 manufactured goods accounted for 99 per cent of all exports, they also accounted for more than half of all imports.

Within this broad trend, Japan continued to suffer a trade deficit with chief suppliers of raw materials, such as Middle Eastern countries, Indonesia, Australia and Malaysia. The Oil Crises of the 1970s taught Japan to diversify its sources of supply as much as possible. On the other hand, North America, South-East Asia and the European Community (EC) together accounted for some three-quarters of Japan's total trade, with each of which Japan enjoyed huge surpluses. In 1989 trade with the USA alone comprised 29 per cent of Japan's trade.

Machinery, especially office and factory equipment and electronics, and motor vehicles were the largest sectors of items traded by value. In 1988, for example, machinery accounted for 44.1 per cent of exports to and 22.6 per cent of imports from the USA, and 50.8 per cent of exports to and 14.8 per cent of imports from the EC; vehicles formed 26.8 per cent of exports to the USA and 18.1 per cent to the EC, and 10.8 per cent of imports from the EC. While machinery likewise accounted for 40–60 per cent of exports to Asian countries, main imports from that region were apparel, machinery, steel and fish.

The chronic trade imbalance and balance of payments surplus rose especially from 1983 and peaked in 1986. As Japan became the world's largest creditor nation, the USA became the worst debtor, and trade friction between the two countries deepened. Complaints against Japan included indignation over the protected market for agricultural produce, accusations of 'dumping' of exports, severe non-tariff barriers such as too-stringent health and saftey regulations, excessive bureaucratic intervention, and difficulty for outsiders to enter the complicated systems of bidding for tenders or distribution of saleable goods. In the 1980s serious issues of intellectual property rights were also raised to prominence in fields of high technology. There is not even agreement over the actual size of the trade imbalance, due to differing methods of data collection, and the *discrepancy* amounted to nearly US$8,000 million in 1988. Even by Japanese accounts, the 1990 surplus of trade with the USA amounted to some $45,000 million, almost as much as the total value of Japan's exports to the EC.

Pressure, mainly from the USA, over issues such as these, resulted in the raising of the value of the yen against the dollar in 1985, and the effect was indeed to reduce the balance of payments surplus. Residual quantitative import restrictions remain only on coal and seventeen items of farm produce (1990), and these are slowly decreasing in number. The complications of the tendering and distributions systems are for the most part not deliberately discriminatory, being just as complicated for the Japanese themselves: foreigners should perhaps not expect preferential treatment, and it is hardly reasonable to demand restructuring of the traditional circulation system to suit foreigner exporters.

Prior to the 1985 rise in the value of the yen, foreign direct investment by Japanese companies was mainly either to exploit cheaper labour in developing countries or to overcome trade friction in export-destination countries. Since then, however, the emphasis has shifted to assembly of goods overseas chiefly for reimporting into Japan. Also, the increased volume of imports from NIEs gave rise to Japanese accusations of 'dumping' of Korean knitwear, with the result that Korea introduced a three-year voluntary export restraint on knitwear to Japan from 1989.

One further important aspect of the balance of payments surplus concerns friction over defence spending. To pre-empt possible rearmament by Japan immediately after the Second World War, the American Occupation Government promulgated a new constitution, Article 9 of which renounces war other than in self-defence. To date (1991), the Japanese people remain firmly pacifist. In 1976 the Cabinet fixed a ceiling of 1 per cent of GNP on defence spending. There has been considerable pressure from the USA to raise the limit on defence spending, partially to soak up some of Japan's surplus and partly to relieve the USA of some of the burden of its security commitment in the Pacific region. So far, the Japanese have not conceded. The outbreak of the Gulf War in 1991 served to intensify this friction.

## AID

Japan has faced no less criticism over its stance on aid to developing economies. Japan became a member of the Organization for Economic Co-operation and Development (OECD) in 1964, including its Development Assistance Committee (DAC). DAC sets 0.7 per cent of GNP as its guideline for a desirable level of Official Development Assistance (ODA) contributions. In 1988 only four of the eighteen member nations met this guideline, and Japan was not one of them, ranking only eleventh in terms of contributions as a proportion of GNP, at 0.32 per cent; this was on a par with the UK. Moreover, the grant element was low, with the average for DAC member countries being 84.3 per cent in 1987, while Japan's was only 47.3 per cent. However, Japan's actual contributions to ODA were US$9,134 million in 1988, second after the United States.

Since 1985 Japan's total contributions to aid, including those other than ODA, have been the highest in the world. This is largely on account of the rapid rise in foreign direct investment by the private sector, which in 1987 ($7,421 million) was almost on a par with ODA contributions ($7,454 million) for that year. As a result, the total value of aid rose by 40.4 per cent over 1986, and accounted for 0.86 per cent of GNP ($20,462 million).

## CULTURE: DIVERSITY IN HOMOGENEITY

Japan's large and dense population is generally regarded by the Japanese themselves no less than others as being homogeneous. Racially they are Mongoloids, and broadly speaking their culture is distinctive from all others and similar over the whole country. There is only one working language, Japanese, and one national costume, the *kimono*.

However, just as there are many dialects to the language, there are innumerable regional and local variations in customs such as cuisine, festivals, handicrafts and agricultural techniques. There is a great diversity of facial and stature types, and recent research by physical anthropologists suggests the admixture some two thousand years ago of South-East Asian type stock with elements from continental Asia; how precisely this occurred remains uncertain. Clearly, Japanese culture is syncretic, owing much to a variety of sources. The country's political isolation from the rest of the world from the mid-seventeenth to mid-nineteenth centuries was also formative of Japanese cultural identity.

The animistic and pantheistic religion indigenous to Japan known as Shinto still thrives, along with Mahayana Buddhism, which was imported from Korea around the sixth century AD. Chinese Daoism was an important early influence on folklore and superstition, and Chinese Confucianism remains a major philosophical influence on social organization, notably the strictly hierarchical structure of Japanese companies and sports clubs and the subordinate status of women. These religious and philosphical beliefs are not considered mutually exclusive and are practised by all Japanese to some extent.

The changing position of women in Japanese society is partially reflected in their participation in wage labour. While 8.0 million women were engaged in wage labour in 1962, the figure had doubled to 16.6 million by 1988. Whereas 55.3 per cent in 1962 were 'never married', that is mostly aged under 30 years, 67.6 per cent by 1988 were either married, separated, divorced or widowed, and the overwhelming majority of these were aged 35–54 years. This trend reflects both higher participation rates for young women in higher education rather than employment and greater social acceptability for women to remain in or re-enter the labour force after marriage.

Despite much public debate in the 1980s on the emergence of 'career women' in management positions, they constituted only around 1.0 per cent of employed women in 1989. Generally, women are disadvantaged in wages, other financial allowances, conditions of employment and prospects for promotion compared with male counterparts, and incomes average only 60 per cent of those for males. Very few are engaged in the 'lifetime' employment system prevalent in the managerial and executive white-collar sector of manufacturing and service industries and the civil service. However, there is a stark contrast between the 'public' image and 'private' reality: in the home, Japanese women typically reign supreme with a degree of unchallenged freedom rare in Western societies.

It is less well known outside Japan that within the largely homogeneous Japanese population there are in fact three minority groups, albeit small, all of which are more or less disadvantaged. In Hokkaido there is a very small population of indigenous inhabitants, known as Ainu, who are both ethnically and linguistically different from the majority population. The Ainu were largely overwhelmed by Japanese colonization of Hokkaido in the mid- to late-nineteenth century. The 1980s saw a reappraisal of Ainu ethnicity and a revivalist movement for the Ainu language.

Another minority group is termed Burakumin. Ethnically they are indistinguishable from majority Japanese, but they are nevertheless a distinct minority, being outcastes. Traditionally relegated to 'polluting' occupations such as butchering, tanning and leather-work, they were officially emancipated in 1871. However, discrimination against them in matters of marriage and employment persists. Since 1969 the government has invested considerable sums in upgrading infrastructure and living conditions in outcaste settlements and in anti-discrimination education among the general public. In 1986 4,603 settlements containing a population of 1.16 million were eligible for such special funding, although actual figures are no doubt slightly higher than official figures suggest. Nearly 80 per cent were in western Japan, and with continued urban sprawl some 60 per cent were mingled with 'ordinary' residential areas.

A more recent minority group comprises non-Japanese nationals, who numbered approximately 890,000 in 1989. Most of these (76 per cent) were Korean, particularly Koreans who were forcibly transferred to Japan earlier in the twentieth century or their Japan-born descendants. The conditions for the granting of Japanese citizenship are best described as ungenerous, and mandatory finger-printing of all long-term foreign residents was frequently chal-lenged during the 1980s as a violation of human rights. With increased 'internationalization', such discriminatory practices are gradually being relaxed. 'Internationalization' became a national catchphrase from around 1980, accompanying the emergence of Tokyo as a leading international money market. It is used not only to denote increasing acceptance of equal opportunities for non-Japanese residents in the country but also to advocate greater participation and leadership in international and global matters by Japan.

## SUMMARY

From around 1964 Japan came to be categorized as a 'developed' nation, and since then it has come to resemble Western countries, in terms of various geographical indicators, more closely than it resembles its Asian neighbours. The population aged rapidly and became increasingly urbanized. The importance of agriculture declined markedly and was replaced by manufacturing industries. During the late 1950s and 1960s, there was rapid economic growth based on heavy industries, and industrialization concentrated into the Pacific Coastal Belt. This gave rise to pollution and congestion on the one hand and rural depopulation on the other.

Due to pollution, congestion and increased costs for raw materials and labour in the 1970s, the relative importance of heavy industries declined, while production of high value-added, know-ledge intensive products such as electronics expanded greatly. An enormous trade surplus contributed to both increased prominence of Tokyo as a leading world financial market and increased trade friction during the 1980s. From the mid-1980s the economy was restructured so as to be led by domestic demand rather than exports. Stable economic growth continued throughout the 1980s, accompanied by further concentration of population and capital into the Pacific Coastal Belt and the increasing importance of the tertiary sector.

# NOTE

1 DID: defined as contiguous census districts of 5,000 persons or more with a population density of 4,000 per sq km; this figure was revised for the 1985 census to 5,000 per sq km.

# REFERENCES

Keizai Kikaku-cho (ed.) (1990) *Heisei 2 nen-ban Keizai Hakusho*, Tokyo: Okura-sho Insatsu-kyoku.

Soga Tetsuo (ed.) (1985) *Nihon Dai Hyakka Zensho*, 25 vols, Tokyo: Shogakkan.

Statistics Bureau, Management and Co-ordination Agency (annually) *Japan Statistical Yearbook*, Tokyo: Japan Statistical Association and Mainichi Newspapers.

Takahashi, N. (1990) 'Recent trends in the activities of financial institutions in Tokyo', *Geographical Review of Japan* 63(1): 25–33.

Uchida, Y. (ed.) (1990) *Ima 'Nihon' ga Osen sarete iru*, Utan 'Kyoi no Kagaku' Series 3, Tokyo: Gakken.

Ueda, J. (ed.) (1991) *Chiezo 1991*, Tokyo: Asahi Shinbunsha.

Umeda, A. (ed.) (1987) *Tokyo: A Bilingual Atlas*, Tokyo: Airisusha.

Yamashita, S. (1990) 'The urban climate of Tokyo', *Geographical Review of Japan* 63(1): 98–107.

Yano Tsuneta Kinenkai (ed.) (annually) *Nihon Kokusei Zue*, Tokyo: Kokuseisha.

Yano Tsuneta Kinenkai (ed.) (1990) *Deta de miru Kensei 1991–92 nen-ban*, Tokyo: Kokuseisha.

# 8

# THE CHANGING GEOGRAPHY OF SIBERIA, CENTRAL ASIA AND MONGOLIA

*Tamara Dragadze*

## INTRODUCTION

In a seminal paper presented in 1904 that influenced geopolitical thinking for decades after, and which is still worth reading for some of its insights, Halford Mackinder divided the world's land masses into three zones, which were arranged more or less concentrically. The inner of the three was called the Pivot Area, and included all the area that this chapter considers (except for the Far East of the former Soviet Union which constitutes the coastal parts of the north-west Pacific). This area was deemed wholly continental (see Figure 8.1). Around it Europe, the Middle East, and the rest of South, South-East and East Asia comprised the Inner Marginal Crescent, partly oceanic and partly continental. The rest of the world including the Americas, Africa and Australasia was the Outer or Insular Crescent, deemed wholly oceanic. By the terms oceanic and continental he was talking mostly about the possibilities of movement and transport, and by implication the means for military control and economic exploitation. The Pivot was the Pivot of History, the source of the Asiatic hordes who had from time to time invaded the lands of the Inner Crescent. It is a region which is the most landlocked on earth, if one discounts navigation of the Arctic Sea. It is also the most continental in climatic terms, being most insulated from maritime tempering, and having the most extreme seasonal temperature changes, and low levels of precipitation. Although in the south (Central Asia) it has harboured great and powerful civilizations, for the most part settlement densities have been very low throughout history.

The fact that the area has been identified in this singular manner should not obscure the very great physical differences that occur within it. These are mostly the result of a combination of two major factors – the differences between mountains and plains (steppes) and the differences in latitude that distinguish between the permafrost of the northern tundra and the (summer) heat of the southern deserts, passing in between through belts of taiga (coniferous forest) and prairie-like grassland. However, through all these differences runs one constant theme – that of the fragility of the major ecosystems. In the tundra where the growing season is very short and species diversity low, the mantle of vegetation and soil, much of it over permafrost, can easily be disturbed yet will be replaced extremely slowly. Montane soils are everywhere thin and easily eroded, but even more so in arid areas. The great grasslands of the steppes may appear to be stable, but the precipitation is low enough that if they are ploughed the risk of wind erosion is very high. Lake Baikal in Siberia contains one-fifth of all the world's fresh water, because of its great depth. It has a unique ecology, and has by far the largest species diversity of both flora and fauna of any fresh-water body on earth. But the ecology is fragile. For part of the year frozen at the surface, rates of biological action are low, and, given the vastness of the water body, the ability of the ecosystem to absorb pollution is also low.

Because these environments are in many ways unattractive for human settlement, the cultures that have been present here through history have been self-reliant, low density, and careful in the manner of exploitation of local resources. Most have been nomadic, diluting their impact over wide areas. It is only in the south, in Soviet Central Asia, that these considerations do not apply, where the waters flowing from the Hindu Khush, the Pamirs and the Tian Shan have provided for irrigated agriculture and settled cities for millennia as they reach the sunny and hot (in summer) plains.

Conventionally the whole region is Asiatic, but the largest part of it, Siberia, is part of European-dominated Russia (until 1991 the Russian Soviet Socialist Federal Republic (RSSFR)), a constituent of the former Soviet Union. Siberia, whose land area is almost half that of the whole of Africa, is settled even now by only 40 million people. In the west it was settled by the Tatars. In the east and north mostly by people related to the Eskimos of other arctic regions, but their numbers have always been few. In 1581 the Cossacks defeated the Tatars near their capital of Sibir (east of the Urals and close to

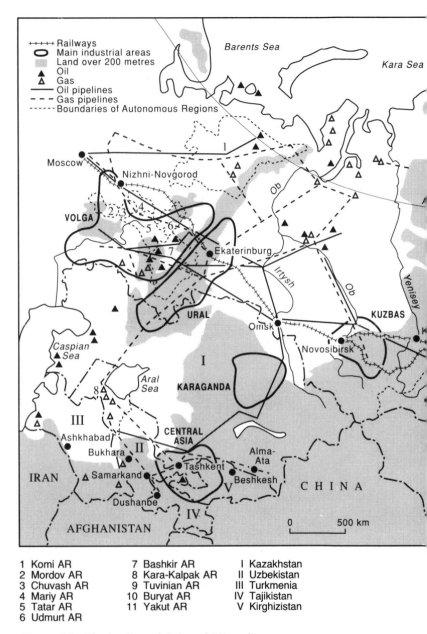

Legend:
- ┼┼┼┼ Railways
- ⭕ Main industrial areas
- Land over 200 metres
- ▲ Oil
- △ Gas
- —— Oil pipelines
- - - - Gas pipelines
- ········ Boundaries of Autonomous Regions

Barents Sea

Kara Sea

Moscow

Nizhni-Novgorod

VOLGA

Ekaterinburg

URAL

Omsk

KUZBAS

Novosibirsk

Caspian Sea

Aral Sea

Ob

Irtysh

Ob

Yenisey

I KARAGANDA

III

Ashkhabad

Bukhara

II

CENTRAL ASIA

Samarkand

Tashkent

Alma-Ata

Beshkesh

V

Dushanbe

IV

IRAN

AFGHANISTAN

CHINA

0    500 km

| 1 Komi AR | 7 Bashkir AR | I Kazakhstan |
| 2 Mordov AR | 8 Kara-Kalpak AR | II Uzbekistan |
| 3 Chuvash AR | 9 Tuvinian AR | III Turkmenia |
| 4 Mariy AR | 10 Buryat AR | IV Tajikistan |
| 5 Tatar AR | 11 Yakut AR | V Kirghizistan |
| 6 Udmurt AR | | |

*Figure 8.1*  Siberia, Central Asia and Mongolia

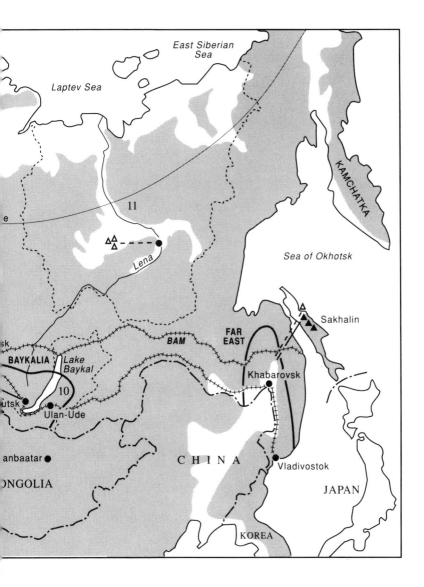

East Siberian
Sea

Laptev Sea

KAMCHATKA

11

Sea of Okhotsk

Lena

Sakhalin

BAM
FAR
EAST

BAYKALIA  Lake
Baykal

10

Khabarovsk

sk

utsk
Ulan-Ude

C H I N A

Vladivostok

anbaatar

JAPAN

NGOLIA

KOREA

contemporary Tobolsk, and from whence the name Siberia), and opened the way for penetration into the interior. Starting in the seventeenth century, settlers from the European areas of the newly found Russian Empire began in larger numbers to infiltrate the area to exploit its resources – which mostly meant fur in the early days. The move paralleled the trappers colonizing the Canadian north. The chances of the Tatars, Yakuts and Buryats of establishing independence now for Siberia are slender, but greater than the chances of the Eskimos reclaiming Canada from the Anglo-Saxons and French. Siberia and the Far East have thus become not just colonized Asiatic domains of imperialist or communist Russia, they are currently incorporated, and possibly permanently so, into the 'European' fold. The Czarist government also began the economic and military integration of at least some parts of Siberia, by building the Trans-Siberian Railway. In the era just before and after the Second World War, there were intensive efforts to develop the resources of the area – mostly in the mining of coal and the extraction and processing of metal ores to supply new heavy industries. These efforts were in part rewarded by the Soviet Union's continued ability to fight Nazi invasion even after most of European Russia had been overrun, a scenario foreshadowed by Mackinder's writings.

Newly independent Kazakhstan, territorially as big as the whole of the EC countries, and the Republics of Central Asia have strong cultural identities of their own, and a settlement history dating back millennia. In Roman times the Silk Route that brought silks from China to the Mediterranean ran through the region. When Islam spread to the area this became only one more feature that distinguished it clearly from Russia. But the process of conquest and incorporation by the Czars and then by the Soviets had by the 1920s finally incorporated the area as republics of the USSR. Thus this area, however, is not part of Russia, and despite large-scale Russian settlement in the past four decades, has again established separate political and national indentities.

In reading what follows about recent changes in Siberia, Central Asia and Mongolia it must be remembered that in many areas, there is no settlement and no development – though this does not guarantee pristine conditions free from all pollution. But settlement is sporadic, cities or clusters of towns and villages isolated from others by many miles of poor roads or creaking railways, if connected directly at all. Much of the mineral wealth of the area is still unknown, and many of the known resources unexploitable, unreached by communications and in any event economically unattractive because of the haulage involved. Much of the development that has occurred has been possible because of the powerful authority of the Czarist government, or the command system of the former Soviet central government. In the future it is not certain that an equivalent determination will overrule the economic objections. This does not mean that specific resources will not be tapped, but that attempts to establish viable local urban economies may falter.

In the last twenty-five years, these areas have been dominated by a particular political ideology which set inflexible patterns of economic practice. The next twenty-five years, however, might perhaps witness infinitely more radical change since it is only in 1990 that attempts at fundamental reorganization of agriculture and industry have been envisaged. More importantly, though, changes in the political structure of the area not only could have strong economic implications as a whole but also might alter the patterns of dependency of the whole area on Moscow, not only in these regions of the former Soviet Union but in Outer Mongolia as well. Dependency on the rouble and other shared economic factors will continue to give Moscow predominance for a while yet.

In December 1991 the three Slavic Republics of Russia, Ukraine, and Byelorussia, having previously declared independence from the Soviet Union, signed an agreement in Minsk for the formation of a Commonwealth of Independent States. The Central Asian and other republics were surprised by their exclusion from this event. Although eleven of the former fifteen republics joined at a later date, the latecomers are unlikely to forget it, and some will seek other forms of alliance. The Commonwealth has had a precarious start. In January 1992 all the republics recognised implicitly the dominance of Russia as they had to react to the latter's removal of price subsidies. Over the next twenty-five years policies designed to increase independence can be expected – very different from the policies of the last twenty-five years. The overall policy of the USSR was to 'integrate' all the republics through economic interdependence with the Centre. The reflection of efficiency was not characteristic of Soviet administrative thinking and designations, where the interests of the Communist Party apparatus had supremacy. Thus the designation and management of administrative regions to achieve this integration have not proved economically efficient. For example, some so-called 'economic regions' which should share the same administration because they share the same kind of natural resources and economic development and might even be adjacent have absolutely no regional political framework to match. The existence of Union republics, such as those of Central Asia, could warrant recognition of republican political identity which overruled shared economic interests. But within the vast Russian Federation, even in the large area known as 'Asian Russia', it was sometimes the case that in two adjacent areas, different branches of the same industry were run separately, the social services were run together and the political administration did not coincide at all. I predict that it is in this field that much of the future restructuring of the administration will take place and that might result in new internal boundaries for some of the general regions. With the rapid growth of regionalism and nationalism, the future political and economic map of the former Union republics might be different. The People's Republic of Mongolia has been only nominally an independent state until the last few years, when, like the former Warsaw Pact countries in Eastern Europe, although less rapidly, it has been able to reassert its sovereign freedoms.

## SIBERIA

Siberia is divided into west and east, covering approximately a surface of 2,427 thousand sq km and 4,122 thousand sq km respectively. Besides being divided formally into east and west, Siberia contains a number of administrative regions, some based around a main town, others reflecting the identity of some of the indigenous peoples who inhabit the area. The Buryats, for example, a people close to their Mongol neighbours on the other side of the former Soviet border, have an administrative region, known as an autonomous republic, within Russia which nominally reflects their identity as a separate people. And, as we shall discuss later, likewise some of the peoples of the Far North, such as the Eskimos who have kinship links with the Alaskans, have to a large extent retained their ethnic identity and parts of their traditional economy. The whole of the area, however, is better known, on the whole, for the pioneering spirit of the Russians and others who moved there to develop agriculture and industry.

### Transport

The Trans-Siberian Railway proper was built to link towns already established by the Russian settlers from one end of the expanse to the other, to the east of Bashkiria (in Cheliabinsk) and in Vladivostok on the Pacific Coast. Work began in 1892 and was completed in 1916. Other railways have been added to complement this main network, although some have not been electrified to this day. As the French geographer Jean Radvanyi has pointed out, the west–east Trans-Siberian Railway concentrates most of the population and most activities along its route at the points where it crosses the north-flowing rivers. Thus, the city of Omsk lies at the crossing of the Irtysh River, Novosibirsk and Barnaul on the Ob River, the large city of Krasnoyarsk is on the Yenisei and Irkutsk is on the Angara River. In cities from Omsk to beyond Baikal, 78 per cent of all western Siberia's population and 85 per cent of eastern Siberia's population are settled in this zone where at least two-thirds of industrial activity and most agricultural activity of the region take place. Although railways are not modernized and efficient when compared with Western standards, they are crucial since the alternatives to rail in Siberia are problematic. Roads to link the different parts of Siberia are notoriously bad, as they are throughout the former Soviet Union. Air transport is used regularly by passengers, although it is known that sometimes they have had to go to Moscow to change planes to get from one part of Siberia to the other. Until recently the rivers and lakes of Siberia were perceived by the Soviet authorities as being important in particular for harnessing hydroelectric energy and they partly determined the location for the development of industry and hence transport.

These patterns were fixed in the Czarist and Stalinist eras, and what is most apparent about the last twenty-five years is the extent to which there is little new development, except for the exploitation of oil and the construction of one major new railway. The building of the BAM (Baikal–Amurski Magistral) has been the centre of many recent polemical discussions. It was started in 1972, opened with great pomp in 1984 but not in service until 1989, and full utilization will not be possible till 1993 at the earliest. The cost of building it has been enormous, much of it in permafrost and in mountain areas, even when young volunteers from all over the former Soviet Union have provided part of the labour. It is supposed to be a second link between Siberia and the Far East and has strategic significance as a line further removed from the Chinese border than the original Trans-Siberian railway. It will be only in the next twenty-five years, however, that its real contribution to development can be assessed, whether new centres of population and industry will be created because of it. The project typifies the relationship between Moscow and the periphery. On the one hand it was the central project of the Tenth Five Year Plan (Symons 1990), showing how much importance Moscow could place on such a venture in a far-flung region. On the other hand, its appropriateness for the region is highly questionable.

## Climate and agriculture

In most of Siberia the continental climate is extreme , with relatively more temperate features in western Siberia. In the southern region of western Siberia there are up to 130 days per year when temperatures rise above 10°C (compared with 100 days per year in eastern Siberia.) But the winters are cruel, for example −17 to −20 C in January, although the temperature falls well below at other times and in eastern Siberia the temperature can fall to around −70°C (average around −40°C). Rain falls at around 400 mm to 600 mm in the west and 300 mm to 400 mm in the east in the summer, which makes crop development difficult. Nevertheless, agriculture has developed in the last twenty-five years despite the mismanagement of collective farming. It is in the next few decades that more decisive changes should take place as policy reforms are implemented.

There are five different kinds of zone relevant for agriculture: first, there is the rich 'black earth', the steppes where soil fertility suggests that yields are potentially high, except for the short growing season and the risk of drought; second, the forest-steppe, grey-forest earths, areas dotted with trees such as in the Altai region; third, the southern Taiga (forest) forming the uninterrupted expanses of lands of 'non-black earth' along the valleys and protected river basins in the area all the way to Baikal; fourth, the rest of the Taiga, which is not arable, and finally, the Tundra, which is not arable. It has been possible to raise cattle and to grow potatoes, cereals and a few varieties of fruit and vegetables (plums and apples, for example) in some parts of the area, but the main problem everywhere is the short growing season. Another problem for agricultural development, however, is the scarcity of labour, since many of the rural areas are sparsely populated (in south-west Siberia, for example, the highest rural density is four inhabitants per sq km). The indigenous populations, such as the Buryats, are more numerous in some of their settlements and if more decision-making power is eventually devolved to them, they may raise their levels of cattle-herding and other similar activities.

There have been some controversial plans concerning both rivers and railways in Siberia. The most hazardous was that mooted on and off since the nineteenth century for the reversal of the flow of the main rivers to point them to links canals to the Aral Sea and even the Caspian, partly harnessing energy but, importantly, for increasing irrigation for arable land. Stalin was enthusiastic about the plan and under Brezhnev it was raised again as a strong project but, thankfully, it was finally dropped by Gorbachev. Part of the reason is not just the cost, but the growing realization that previous massive schemes (like the excessive use of the rivers flowing into the Aral Sea, described below) have incurred horrendous environmental costs. With Russia's commitment to a market economy, the methods and results of financing development in Siberia will change substantially.

## Industry

Industry, in particular the processing of natural resources, grew immensely in the Soviet period. First the Kuzbas region has to be mentioned. In the early eighteenth century this coal-mining area was discovered and mined for local use. In 1930 Stalin ordered the development of this region to the east of the Urals. The most recent statistics show that the 152 million tons of coal mined annually here, with one-third in open cast, represented nearly one third of all coal mining in the Soviet Union. In Kuzbas, heavy industry has been promoted but in steel works in particular. At Kansk-Achinsk the location of lignite so close to the Trans-Siberian Railway gave rise to plans to open fifty pits, the first in 1965, for thermal-electric stations to provide an important source of energy for Siberia which would allow Kuzbas to send even more coal westwards to the European part of the Soviet Union, leaving even less of it for use in Siberia. But both ecological problems and financial ones have been detected and so plans for investment may decline and Krasnoyarsk will have to rely on its traditional industries instead of the new ones envisaged in the Kansk-Achinsk project.

In 1989 the workers of the Kuzbas region shook the Soviet political establishment to its foundations. Always considered loyal to the Communist leadership which, at least in words, had always lauded the importance of the industrial working classes, the miners went on a bitter strike, deploring their living conditions and the centralized management policy which not only resulted in waste and inefficiency, but also favoured other regions in the European part of the Soviet Union for political reasons when the Kuzbas presented more economic promise. Were the demands of the striking miners ever to be met, the whole economy of the region would change, since they favoured not only radical modernization of their industry but also much more logical and profitable redirecting of resources and production. As elsewhere in the former Soviet Union, also, there were demands that more profits be invested in the region rather than expropriated by Moscow to be distributed according to whim anywhere and to anyone. The city of Tomsk, for example, has already benefited from Kuzbas since it became a centre for chemical and petrochemical industry and grew by 1 per cent per year from 1970 to 1989, whereas little growth has been registered in other similar towns. In January 1992, however, the Kuzbas miners were still unsure about the size of the quota of compulsory sales of coal to the Russian central authorities. The reformist Russian government under Yeltsin had promised them the status of a free economic zone – but quotas remain sufficiently high that the miners are still denied any real hope of improvements in safety and living standards.

Novosibirsk, one of the most important towns of Siberia, surrounded by good agricultural lands and the home of machine tool and other similar industry, became home to a new Academy of Sciences. On the outskirts of the city a new town Akademgorod (Academic City) was home to a group of sixty institutes and their scientific staff, in the hope that such a dense academic community in the middle of Siberian woods could propose programmes for the development of west and east Siberia. In recent years there is less secrecy surrounding scientific work and so scholars are able to meet more freely with colleagues from abroad. On the other hand, it has yet to be seen whether any of the plans elaborated in academic institutes will serve the Siberian economy any better than they would in other countries!

The Siberian mountains of the Altai and Sayan might be mineral rich, but significant industrial development has not taken place there as yet. This region therefore represents the predicament of so much of the former Soviet industrial potential. Despite the predilections of their predecessors, policy-makers realize there is a need to scale down the promotion of heavy industry in the area as a whole, yet medium and light industry have not been developed either. It is here that, if one takes an optimistic view, there are likely to be changes in the next few decades.

## Lake Baikal

This is sufficiently important to deserve a small section to itself. The lake has the same surface area as Belgium, and is a mile deep. It has been a genetic generator – spawning more species than any other water-body on earth. It is the sole habitat of 600 forms of plant life, and 1,200 forms of animal life – and in all, two-thirds of all fresh-water species on the planet. As examples, it is the home of 80 species out of 120 species of mollusc and 252 out of 450 crustaceans. Upstream of the lake are the 600,000 inhabitants of Irkutsk, who rely on the lake in part for sources of protein, in particular the omul, a local fish. But this has of late been diminishing in size and the catches diminishing in quantity (*The Times* 8 July 1991).

On the southern shore of the lake at Baikalsk, there are two paper and cellulose factories, planned from the 1950s, and now operational for nearly thirty years. Such mills have peculiarly noxious effluents, which for many years were not properly treated. Now the effluent is treated as well as is possible with current technology, but it is still polluting the lake. Calculations of the economic costs and benefits of such a case are notoriously difficult, and in any event can be immediately discounted because it is impossible to cost now the loss of any species, which narrows future options indefinitely. But according to one calculation, which values the fresh water of the lake at a very low and nominal level, the costs of damage inflicted in three days are worth the total annual output of the factories. This may be hyperbole, but it is nevertheless symptomatic of the fact that there is an aquatic equivalent here of the rain forests – and perhaps an example where a kind of swapping of 'debt' (or assistance) for 'nature' could be applicable.

## The Far North and oil

The Far North of Siberia, around the Arctic Circle, which usually refers to the area north and just immediately south of 70°N latitude, takes in regions of western and eastern Siberia and even parts of the former 'Soviet Far East'. This area contains primary resources of such magnitude that they are as yet not fully known. From the 1930s prisoners of the Gulag camps were used to mine gold, diamonds and nickel. In the 1960s, however, a complete change in attitude to the region took place and it became the focus of the greatest pioneering programme of the period. Siberia has become the chief supplier of gas and oil in the country, mostly from the plains of the Ob River. Largely because of these developments the USSR had become the largest oil producer in the world by 1974. Centres were also developed for the mining of non-ferrous metals of great variety, and the population in the region has increased four times in the last three decades. Whole new towns have grown up. Transport routes, some open all year, some iced up in winter, have been built. It was not till 1985, though, that attention was paid to infrastructures and to the fact that housing was in a catastrophic condition. Planners are having second thoughts, now, about the suitability of gigantic plans, even with American and Japanese aid. Recently they have decided they should scrap a large plan for five new petrochemical centres which, for them to be viable and to have the proper infrastructure, would have required a population of around 1 million. Further exploration is not being supported unconditionally either, even though some resources currently under exploitation, such as oil, are peaking in some areas. New thinking about the kind of labour to attract is also being considered: in the future it might be more profitable to attract more qualified employees than in the past, but better conditions will have to be offered to them. Above all, though, more efficient extraction and distribution of existing resources should be achieved first, it is thought, before embarking on new projects which might have little practical value in the international markets to which this part of Russia must aspire.

The incursions have had an adverse effect on the autochthonous peoples of the north – the Chukchee, the Mansi, Khansi, Evenki and the Eskimos, for example – who saw in the 1960s and 1970s a dramatic reversal of fortunes in a state which had until then relatively favoured them compared to other Soviet minorities. Even relatively larger groups such as the Yakuts have been adversely affected. The whole basis of their way of life was undermined: they lost grazing and hunting lands vital to their reindeer herding, their life expectancy fell and infant mortality rose. Hindered by prejudice and poor education, they have not been absorbed into the new industrial activities and urban life either, run by newcomer settlers who are mainly Slavs (nearly all Russians, with some Russified Ukrainians and Byelorussians). Heartened by efforts made by autochthonous peoples in Canada and Alaska, in the past year or two those of Siberia have also attempted to make their voices heard but it is too early to tell whether there will be any significant changes in the next few years to their plight. The fact that decision-making still lies in Moscow, notwithstanding by a 'democratic' government, gives sustenance to local protest movements.

The ecological impact of the industrial age on the fragile Arctic seas and atmosphere is now internationally recognized. In 1991 seven northern polar nations including the USA, Canada and the former Soviet Union agreed to new international monitoring of pollutant release in water bodies, and atmospheric pollutant release. It is the first stage in co-operation in the more careful development of these lands.

## THE FAR EAST

The Russian state declared an interest in the Pacific coast as early as 1648, with the foundation of Okhotsk (Hunting Place). Eager to obtain furs, the Russian traders had pushed ever eastwards until they reached the sea. Later the strategic importance of having a stronghold along the Pacific, from the Bering Straits to Japan, increased rapidly. Commercial interests were important before the Russian Revolution but economic justification for developing the coast can be assessed only in the future. By current convention developed during the Soviet period, the region known as the Far East includes not only the coastal border and off-shore islands, but also the area inland of the Yakut Autonomous Region, along the Lena River. Many of its frontiers are disputed by the neighbours, China and Japan. There have been skirmishes along the Ussuri and Amur River borders with the Chinese. In the last year or two a commission is working to solve 'finally' the border problems, in the wake of the renewed warming of Sino-Soviet relations. The dispute with Japan, especially over the Kurile Islands seized after the Second World War by the Soviets, has yet to be settled, and continues to bedevil relations.

The region has the least arable area of land in the former Soviet Union, although some herding and a little agriculture in some scattered areas have been developed. In the southern area near Sakhaline and Vladivostok, even rice and soya can be grown. The most difficult problem of the region, however, is that little more than 30 per cent of food needs can be met locally. As everywhere else in the former Soviet Union, mismanagement is largely to blame but this could be rectified and, with intelligent selection schemes, the potential could be developed and the shortages rectified.

The principal cities of the Far East have grown rapidly in the last thirty years, though less so in the last ten. As in Siberia, they have been developed mainly in order to expand industry around the extraction of natural resources. These, however, are usually not fully processed locally and are transported at great cost, usually westwards, to be processed. Local energy resources have yet to be developed and again, the high costs can hardly be justified. Planners in the Moscow ministries during the Brezhnev period decreed that, for example, machine-building or tool-making factories should be built whose output are neither used locally, nor can be sold in Pacific export markets. They have to be transported, again at great cost, either to the European part of the Soviet Union or to Third World countries who accept them as part of tied bilateral assistance. Some of the coastal cities, of course, have been developed mainly for their military importance. Most important, however, is the fishing industry, and in Vladivostok fish processing for distribution is very important. The ship-building and ship-servicing industries are still important, although in decline. On the island of Sakhaline, the southern half of which belonged to Japan from 1905 to 1945, but now part of Russia, petrol and gas are extracted and sent by pipeline to Komsomolsk inland. It has a varied population, including Koreans deported there during the war, a reindeer herding indigenous population, but as in so many other far-flung regions of former Soviet Asia, an immigrant Russian majority.

The Far East as a whole has about 8 million inhabitants, which is not a large portion of the total 286.7 million in the former Soviet Union as a whole (1989 census figure). According to the 1989 census, Vladivostok has 648,000 inhabitants and grew by 18 per cent in the past decade. Khabarovsk (601,000) grew by 14 per cent in the same period. The much smaller cities of Yujni Sakhalinsk (157,000) and Magadan (152,000) grew by 12.5 and 25 per cent in the past decade. The amount of forced labour from prison populations in this region is not revealed in statistics, although one presumes these have diminished since 1957. Neither has there been much discussion of the previous conditions of people who opted to move to the region. Above all, one must not forget that the central planners had changeable policies. At times they allocated double and sometimes triple wage scales to workers according to the geographical zone they worked in, and most of these areas scored high on the scale. One often reads, however, of harsh conditions. Supplies, particularly food supplies, are sparse and infrastructure undeveloped and there is wide discontent, especially inland. Price rises in 1992 as subsidies are removed can only exacerbate the situation. Thus, despite the attraction of relatively high wages, the population turnover is high. Along the coast the situation is somewhat different, with the attraction of access to some foreign goods smuggled in and the chance of travel some of its industries can offer.

The whole strategy of the region has to be revised and redeveloped in the next few years, if the stagnation and decline of the economy is to be halted. Waste of natural resources will have to be halted, probably through developing only those that are economically viable and by processing them locally. Further sources of energy will have to be developed regionally too. Then the infrastructure, and this includes tackling food shortages as well as housing and communications, will have to be revised and improved. Eventually, too, Russia's Far East will have to compete with the other countries of the region, such as Japan and Korea. It is already signing agreements with many of the countries of the Pacific region, such as Australia and New Zealand. There is a strong suggestion that certain coastal areas should become free economic zones, in the hope of attracting foreign investment. One can only hope that political stability will permit this to be achieved.

## KAZAKHSTAN AND CENTRAL ASIA

Although the indigenous peoples of Siberia and the Far East have based their inhabitation of those regions on subsistence and sustainable economies since time immemorial, Central Asia presents a very different historical picture. Being the location of old civilizations, at a trading cross-roads between significant movements of goods, people and ideas, Central Asia had an identity all its own which, despite military conquest by Russia in the nineteenth century, could not be eclipsed by the arrival of Slav 'pioneers' even in the Soviet period. Furthermore, according to the 1989 census data available, the high birth rate of the indigenous population in comparison with that of the Slav settlers has resulted at present in their achieving numerical dominance. The main religion is Islam, but its significance other than cultural should not be exaggerated. Many of the Central Asian peoples were nomads with shamanist rituals where the official nature of Islam had little place. Seven decades of militant atheist programmes encouraged by the government have produced a secularized version of Islam throughout the former Soviet Union today. Furthermore, without the economic and property base for the official Muslim clergy and institutions, even if there is an upsurge of religious fervour it is unlikely to resemble that witnessed in Muslim countries outside the region. In the last few years, however, it is the relative poverty of Central Asia, with its high infant mortality rates, its polluted lands and its high unemployment rates that have been revealed in the era of glasnost. This, rather than religion, was enough to fuel the dissatisfaction with Moscow. With the demise of the USSR and the about turns of Russia, which has shown a reluctance to shoulder any responsibility for poor Central Asian Republics, the latter are casting around for new allies. As Muslims, they have appealed to fellow Muslim countries such as Saudi Arabia, which has so far donated religious books. Pakistan has agreed to expand trade. But the politicians of Central Asia know that any substantial aid, particularly of technology, will have to come from the West.

## Kazakhstan

The inclusion of Kazakhstan with Central Asia as done here is the subject of strong disputes. This is especially so among Russian specialists who have been keen to stress the importance of the lands to the north where predominantly Slavs came to settle. Another reason for setting Kazakhstan apart from the other Central Asian republics is its sheer vastness and immense variety of terrain.

It is a former Union Republic, created by Stalin, covering a surface of 2,700,000 sq km, stretching from the European part of Russia by the Volga delta right up to the Tian Chan mountains by the Chinese border. Parts of this area were colonized by Russians from the eighteenth century onwards, and until very recently, the Russian and Slav element in the population of Kazakhstan formed a majority. The opening of the 'virgin' and 'idle' lands for settlement and the growth of industry in places such as Karaganda provided Russians with the chance to settle there. In 1979 the Kazakhs constituted a mere 36 per cent of the population of around 14.5 million. Partially released 1989 census statistics indicate that the population has grown to 16.5 million, and that a combination of the high Kazakh birth rate and population migration means that perhaps the Kazakhs are again just in a majority in their state, though not in principal urban areas. In any case, the Kazakhs have been established as the dominant political identity for the republic, and in view of riots in the capital, Alma Ata, if we are to observe any changes in the next twenty-five years it will surely witness increased assertions that Kazakhstan is the land of the Kazakhs, even if they will accommodate the non-Kazakh majority into their plans. This will not be unproblematic especially if one remembers that Kazakhs, though vociferous in their protests, form only 16 per cent of the population of their capital. Russian nationalists with extremist views might make territorial claims on Kazakhstan.

There is an increasingly articulate Kazakh intelligentsia although the number of disaffected Kazakh youths is growing too. As elsewhere in the former USSR, literacy in the indigenous language, albeit using a Cyrillic script, is said to be almost complete. There are indeed thirty-two newspapers and periodicals in the Kazakh language. Local periodicals and literature are published in Russian, where they predominate not in the intelligentsia but in industry. Besides the Russians, there are also ethnic Germans and others for whom Stalin used Kazakhstan (and Uzbekistan) as a dumping ground. It is nevertheless for political reasons rather than economic or ethnic calculations that Kazakhstan is included in Central Asia here.

In Kazakhstan, unlike the other republics of Central Asia, heavy industry plays a leading role. Nevertheless, Krushchev launched a grandiose scheme for colonizing the 'Virgin lands' of Kazakhstan for growing cereals on a large scale. This was achieved mainly by the arrival of Russians and Ukrainians. The traditional pastures for nomadic Kazakhs had long been diminished when, in the process of forced collectivization in the 1930s, the native Kazakhs destroyed many of their herds and had to abandon their traditional way of life. Agriculture and herding are important for the Kazakh economy and used to provide a contribution to the food supplies of the Soviet Union as a whole. The poor performance of the 'Virgin Lands' scheme and the political transformation suggest that there will soon be another major change in the region.

The main industries are coal in Karaganda and steel and copper production in Dzhezkazgan, and iron mines are being developed in the region of Kustanai. All these industries were supposed to be integrated with those of other parts of the Soviet Union, especially those of the RSSFR, but feeble transport networks and clumsy administration, as usual, impeded success. Should the devolution of power of decision-making be regionalized and direct agreements between enterprises be developed, then this situation could change.

One significant problem and one that should be monitored in the future, is that the main atomic testing ground of the former Soviet Union is in Kazakhstan. There have been strong protests since 1989 but it has yet to be seen whether public opinion will affect the military to the extent that they will comply fully with demands to abandon nuclear testing, some of it carried out with sinister consequences for the local population. The new Kazakh government has asserted control over nuclear weapons on its territory, although this may not be realistic in view of the protests made by Russia, whose government sees itself as the heir to Soviet power. If Kazakhstan decides to surrender the weapons, it risks being 'blackmailed' at a future date. The future of the Space Centre at Baykonur is also shrouded in confusion.

## Uzbekistan

This Central Asian republic is often seen as the most dominant in the area south of Kazakhstan because of its size and political significance. Tragically, Uzbekistan, which was paraded by Soviet officials as a success story to be emulated by other Third World countries if only they were to adopt Soviet patronage, is instead in the past two years often pointed out as an entire disaster area. The reasons for this stem mainly from the ecological damage that has been done to the soil and to the water resources on the one hand and the rising population, with the threat of large-scale unemployment, on the other. Uzbekistan covers an area of 447,400 sq km and in 1989 had a population of around 16.24 million, a rise from 1979 (14.7 million) of some significance. In parts it has one of the highest densities of rural populations in the former USSR.

Stalin created the entities and boundaries of the Central Asian republics where before there had been a mix of Russian administrative territories, such as that known as 'Turkestan', and the emirates where the Russians exerted a form of indirect rule, such as in the 'Emirate of Bukhara'. These territories had not had clearcut local homogenous ethnic settlements and Stalin's arbitrary acts, such as attributing Bukhara and Samarkand to Uzbekistan rather than to Tajikistan, is still resented by the latter. Where other peoples straddle borders with Uzbekistan, such as the Kirghiz, and where there are large settlements of populations deported and dumped in Uzbekistan such as the Meskhetian Turks, the intensification of anxiety over access to land and water has often spilled over into ethnic conflict ostensibly because of other matters. The capital is

Tashkent, but the several regions of Uzbekistan with their provincial towns, such as Andijan, have fostered local loyalties that sometimes become visible in the kinship and friendship networks for jobs in high places.

As in ancient times, Uzbekistan with the other Central Asian republics has to overcome its shortage of water through irrigation systems and judicious choices of crops. The whole Central Asian area, however, in Soviet times was galvanized to produce cotton, more than 90 per cent of the production of the whole of the USSR, regardless of ecological cost (Carley 1989). The over-fertilization this entailed has resulted in what is seen as the virtual poisoning of the soil and the drinking water supplies in many areas. The misuse of canals and the water system in general has had devastating effects for the whole region, Turkmenia and Uzbekistan in particular.

The Aral Sea was ranked fourth among the world's largest inland lakes, spreading over 68 million sq km, maintained by the flow of the AmuDarya and SyrDarya Rivers. From the mid-1960s incompetently thought-out new irrigation canal systems extracted so much water from these rivers that 65 per cent of the incipient volume has been lost (Nasar 1989; Oreshkin 1991). The depth of the Aral Sea has dropped by some 15 metres, and the surface area of the water body shrunk to half. Around 3 million hectares of the sea bed have been exposed and have become the site for an active salt and dust release into the atmosphere. In 1987 the President of the Uzbek Academy of Sciences, Professor Habibulaev, stated that between 15 million and 75 million tons of salt annually are being carried and deposited on fields and cities of Central Asia. Around 80 per cent of wildlife has been affected, mostly destroyed completely. Health hazards have increased dramatically. Infant mortality is at least two to three times higher than the average for the USSR as a whole, with for example infant deaths from hepatitis being 526,000 (in Karakalpak in Uzbekistan) in 1988 compared with the 251,000 average for the USSR.

More pernicious, even, than the Aral Sea disaster are the effects of over-fertilization. In Uzbekistan the growth of supplies of commercial fertilizers to the region was as follows: from 345,000 tons in 1960 and 729,000 tons in 1970 to 1,288,000 tons in 1985, and yet cotton production, having reached a peak in the 1970s, has declined ever since. Health care is rudimentary but ubiquitous, and yet infant mortality in Uzbekistan for 1988 is above 46 per thousand, compared for example to 11.6 per thousand in Lithuania, with some

Uzbek regions reporting one death among every ten infants. It is difficult to know what the situation was like before the Gorbachev era when such secrecy surrounded these issues, yet it is recognized that certain viruses of hepatitis and high levels of anaemia were nevertheless not so prevalent as today. In Karakalpak, within Uzbekistan, with its northern border along the Aral Sea, the provision of medical services is said to be particularly low (which compounds the high infant death rate from hepatitis cited above).

Embarrassed by this state of affairs, the Moscow government at first presented explanations for the appalling conditions by blaming the local Uzbek officials whom they accused of being corrupt and local Uzbek customs for being primitive. The backlash against this attutide has resulted in a local assertive nationalism, coupled with the demand that the central authorities take responsibility for the havoc that their planning has caused. To this end Uzbekistan, like the other Central Asian republics, is compelled to insist on joint arrangements with Russia in order for reconstruction and reparation to take place to alleviate some of the hardships the Uzbeks are experiencing today. Their request that Siberian waters be used to replenish the Aral Sea, however, is being resisted at present by the Russians.

## Turkmenia

Turkmenia also has a border along the ill-fated Aral Sea and often its statistics are presented together with those of Uzbekistan in the specialist literature. It is one of the earliest areas in the region to be exploited by the Russians for industrial development resulting from the discovery of oil and gas deposits. The industrial potential, given the natural resources and the experience of industry, should make the republic particularly promising for development and future prosperity. Turkmenia also has the benefit of having the Caspian Sea on its western border, although once again, the Caspian is showing incipient signs of problems similar to those of the Aral Sea. The territory covers 481,100 sq km and the population is around 3.1 million (2.3 million in 1973). The state is the most arid of any in the former USSR. The soil is used, where irrigated, for growing cotton as elsewhere in Central Asia but also for keeping sheep wherever possible in order to export almost as much wool as from Kirghizia. Fruit, such as melons, are cultivated for export. Silk production is another Central Asian activity present in Turkmenia. Yet poverty, with high rates of infant mortality and malnutrition, have recently been revealed in the wake of the government policy for more open access to information. Curiously, the old Soviet government, anxious about falling birth rates among the European population, always had a pro-natalist policy with 'heroine mothers' awarded medals for having more than ten children. Yet it was not able to have discriminatory policies towards Central Asians whose natal practices, on the contrary, should warrant the encouragement of reducing family size. It is in Ashkhabad, capital of Turkmenia, however, that the USSR's first family planning clinic was set up. Yet one of its senior members has told me that its purpose is not to curb the birth rate itself, which is seen to be beneficial to Turkmenia in its struggle for more independence from Moscow. Instead, its policy is merely to promote better health for mothers and children and to assist in spacing the birth of children if that is deemed necessary for any particular case.

The Turkmen politicians are reasserting their own national identity and, for example, a law concerning Moslem circumcision has recently been passed in the Turkmen parliament. They are aware of their fellow ethnics living in Iran and elsewhere but they are unlikely at present to divert their political attention to them and are instead concentrating on achieving the conditions where their own republic can improve and profit from their natural resources and growing workforce, to eradicate the encroaching poverty and pollution that appears to have acquired alarming proportions.

## Kirghizistan

This is a mountainous republic covering 198,500 sq km except for the important valleys along the Chu and Talas rivers and part of the Fergana valley, in which most agriculture, industry and settlements are placed. It is not the only homeland of the Kirghiz, there being others who live in China. Despite the natural resources of Turkmenia, it is Kirghizistan which holds third place after Uzbekistan and Tajikistan for industrial production in Central Asia and second place after Uzbekistan for agriculture. Oil has been found in southern Kirghizistan, but only after drilling 2,000 metres. Some mineral ore was discovered and settlers from Russia were brought in to force the pace of developing according to the usual Soviet standards. Hydro-electric power was promoted, at the cost of misusing scarce water resources. The republic did not escape the obligation to grow cotton, but the main activity has been sheep farming, particularly in the highlands of central Kirghizistan. Industry developed alongside agricultural activity, such as cigarette factories for the tobacco that is grown there and textile factories for silk and wool. In northern Kirghizistan heavy industry has been established, as one of the 'musts' of Soviet planning. Metallurgy and machine tool plants were built and the capital of Kirghizistan, Beshbek (formerly Frunze), is linked by railway to the rest of the former Soviet Union, but particularly with Issikkul and the valleys of Central Tian Shan. In 1973 there was a population of 3 million which in 1989 had risen to 4.29 million.

Anxiety over land shortages might be one of the reasons which lay behind the violence in the city of Osh in 1989 between Kirghiz and Uzbek inhabitants over the allocation of land by the local government. If economic reform is to be developed in Kirghizistan the social problems accompanying allocations for private use and privatization generally will have to be faced. The pace at which Kirghizistan develops will depend on the versatility, and the unhampered conditions to use it, of the local politicians in developing alternative industries and to overcoming the problems of pollution and water scarcity. With its breathtaking mountain scenery one can expect that eventually a valuable tourist industry could be developed. Since the demise of the USSR, Kirghizistan has adopted the most democratic government policy of the Central Asian Republics so far. It has yet to be seen if there is tangible Western support for these policies.

## Tajikistan

This is territorially the smallest of the Central Asian republics, 143,000 sq km, but it has a larger population than Turkmenia with 4.36 million in 1989 (3 million in 1973). The greatest part of its territory lies above 2,000 metres in altitude. The Pamyr mountains form its southern borders and through them passes one of the main routes for Russians into Afghanistan. Culturally, the Tajiks identify with their Iranian forebears, although they are Sunni Moslems and not Shia, and they attribute Tajik identity to ancient poets and scholars such as Omar Khayam or the physician Avicennus. They see their culture as vastly superior to that of their Russian colonizers who, they assert, brought them technology but little in the way of social manners and civilization. At present Tajiks see their task as attempting to disengage themselves from Moscow economically, in a system where domination and subordination still hold, and to re-establishing links with the Middle Eastern world, and through it with Europe whose technology they crave. They have not escaped the same problems as the rest of Central Asia, however, such as the compulsory growing of cotton. A high birth rate, once praised as patriotism, is now revealing an accompanying high rate of death and disease which the medical services are unable to cope with.

The capital Dushanbe was built in Soviet times, with a majority non-native population. As in the other Central Asian republics, it boasts a theatre, opera house, university and Academy of Sciences. Recently there has been an exodus of settlers and a city majority of Tajiks is expected to develop in the next few decades. As in its neighbouring republics, the poor food distribution and low wages have discouraged Tajiks from moving out of rural areas into towns where they will lack their village kinship and family networks for support.

In the Pamyrs lies the territory of the Mountainous Badakhshan Autonomous Region, with its local capital of Khorog, where plans for extensive hydro-electric development along the River Penj were seen as the way forward for economic development there, but the wisdom of these schemes is now being openly questioned.

At the turn of the century the Tajiks had sympathized with the Pan-Islamic movement that attempted to unite Muslims in the Russian Empire, but would not have participated in attempts at Pan-Turkism, since they are outside that loose ethnic and partially linguistic union, which largely sets them apart from other Central Asians. They also, perhaps more than any other people in the region, deeply resent the arbitrary way in which Stalin drew the borders of the republics which, according to them, deprived them of Samarkand and Bukhara where they claimed to be in a majority. The last figures before the Soviet invasion of Afghanistan showed that there were more Tajiks in Afghanistan than in Tajikistan, but their economic poverty was greater than that of Soviet Tajiks. Although there is hope for intensified cross-border cultural and religious links once that country achieves stability, the Republic of Tajikistan is intent on consolidating its own position and, like the rest of the area, in achieving a balance between benefiting from ties with Moscow from which economic independence is as yet in-conceivable, on the one hand, and achieving on the other hand the degree of sovereignty necessary to retain profits for local investment and to help its people to escape from disease, pollution and land hunger which has blighted the region in recent times. At the end of 1991 former communists defeated the opposition in elections, but not by much. The growing power of the Islamic Renaissance Party which has links with fundamentalists in Afghanistan and Iran may lead to a much more vigorous contest soon.

## MONGOLIA

The People's Republic of Mongolia is very sparsely populated. It extends over 1,565,000 sq km and it has a population of barely 2 million. Traditionally, and still today, herding plays a vital part in the national economy, but now 51 per cent of the population is urban. As in the former Soviet Union, whose form of government it emulated and with whom it had close ties ever since 1921, there has been a major increase in the mining and industrial sectors.

Although not more than 5 per cent of the territory is mountainous, most is formed of various kind of steppe as well as including the Gobi desert. The five ecological zones of Mongolia are mountain (5 per cent), forest steppe (23 per cent), steppe (27 per cent), desert steppe (28 per cent) and the Gobi desert (17 per cent). The variations in temperature are extreme, from $-40°C$ to $+40°C$, with rainfall annually between 100 mm and 300 mm, which is little, especially since this can include winter snow.

Following the Soviet-type communist system, however, the mining and machinery industry was developed mostly for export to the Soviet Union at artificially low prices and with scarce regard to either local control of capital investment or to local ecological needs. Agriculture was collectivized. Yet rudimentary education was spread universally and, in the capital, there was a Soviet-style organization of academic life. A University and a Mongolian Academy of Sciences was established which has in more recent years fostered direct relations with academic communities abroad. The Union of Artists and the Union of Writers produced national elites, controlled by the Communist Party as elsewhere. The political elites were Moscow trained and it should be remembered that the Mongolian script was abolished and replaced by a Cyrillic script as a means of cutting Mongolia from its past, and to win the battle against religion, a task considered very important by the early communists.

In 1989 the winds of change in the Soviet bloc reached the

People's Republic of Mongolia. Students and intellectuals began demanding democratic freedoms for their political system and this has resulted in a government being elected in 1990 in which over 40 per cent are non-communists. The new Mongolian authorities have now begun activities on three fronts: first, the People's Republic of Mongolia seeks to reorientate its trade away from the former Soviet Union and to establish links instead with Japan and other capitalist countries; second, there is a strong ecological movement that is demanding that no further factories be built on Lake Hövsgölnuur, and that a strong anti-pollution policy be applied, in particular against the old Soviet-built enterprises; third, there is a strong national movement to rehabilitate national culture. All primary schools will teach the Mongolian script and it is hoped that by 1994 or 1995 the script will replace the Cyrillic one universally. Likewise, religion and Bhuddism generally are being recognized as being important components of national culture and lamasseries (traditional monasteries) are being reopened and have been freed from persecution.

The areas where we can expect significant change will be in the new trade directions which might include some importation of labour from neighbouring countries such as China (despite the disapproval by the People's Republic of Mongolia at the treatment of their fellow Mongols in China). With modern technology their mining industry and agricultural production are likely to change. The range of consumer goods available and living standards should improve. Some new links with the autonomous republics within Russia, such as Buryatya and Tuva, are likely to increase and the local culture is likely to become more assertive and provide a lead in the region as a whole.

## GENERAL SUMMARY AND EVALUATION

Many changes have been noted in this chapter, some good, some bad, but all reflecting a particular attitude towards what 'development' has meant in the former USSR. The prime target has not been identified as improving an index such as GNP per capita, nor even to improving social equity and justice, since that has been assumed inherent in the system anyway. Given that the political system would, by virtue of social ownership, benefit all, the aim was to use society to develop the resources available to it, primarily through industry, the hammer having priority over the sickle.

Development of resources even without adequate calculation of direct and indirect costs had been assumed inevitably to lead to the utopian future. The reality has been short-term gain, political expediency and prestige impression-management that took precedence over concerns for economic efficiency, human welfare and ecological security. This was as much the case for agriculture, with the opening up of the 'Virgin lands' of Kazakhstan as for the BAM railway line or some of the hydro-electric dams. The tragedy of the Aral Sea and the emerging tragedy of Lake Baikal are part of the cost of this approach.

Soviet ideology demanded that in managing population resources, priority be given to building an urban proletariat. Peasants were always viewed as problematic by Marxists and it was hoped that eventually the countryside, in a sense, could be transformed for management and for the development of class consciousness among peasants, into a form of industrial complex. Collectivization of agriculture took place for a variety of reasons but they included the wish not only to control the peasantry but also to ensure that the industrial population would be fed. On the other hand, the state took upon itself to be responsible for social welfare, however rudimentary in the spheres of medicine and education. Again the reality does not match expectations. When one takes living standards, which include indices of health conditions, life expectancy, nutrition, literacy and so on, then the area covered in this chapter comes out relatively well in comparison with many Third World countries but badly when compared to the Western standards to which the Soviet Union aspired. The pharmaceutical industry in the Soviet Union was given scandalously low priority. Yet the structures are in place, which, if overhauled and improved, could provide a more adequate medical service in the future.

Compared to most of the Third World, the Soviet system did eradicate illiteracy. Although education was hindered by an inordinate amount of time being allocated to learning official ideology, it was aspired to by all, at the cost of great disruptions to some of the people. One of the main complaints even today of the indigenous peoples of Siberia is that to attend school, their children are often obliged to live in boarding schools from the age of 7. A disadvantage to the ruling authorities, of course, is that a generally high level of education gives rise to dissatisfaction because of high expectations. It should not be forgotten that, bar a few villages with picturesque wooden houses, the majority of the population of

Siberia lives in high-rise flats and considers itself to be urban and modernized. On the other hand, the level of industrial skills is deemed by visiting Western specialists to be generally very low. Retraining on a large scale will be necessary for former Soviet industry to modernize sufficiently.

Much of the future of the whole region depends on several factors which, at the time of writing, cannot be guessed at clearly at all. How will the republics of the former Soviet Union and Mongolia abolish central planning and establish a market economy? Will this include establishing full privatization of land? Will these countries, through having a convertible currency, forge trade links and industrial development with the advanced economies of the OECD? How will the Central Asian Republics negotiate a new form of economic union within the new CIS, or will all half-way houses collapse, and lead to a new 'Balkanization'? Would that depress or improve local conditions? Would the Russian settlers emigrate from these areas? Finally, will open debate permit some of the ecological follies to be rectified and others to be avoided? Although most of the elements of the changing geography of the future will be political and economic, there might be some physical ones too for rivers, lakes and vegetation.

## Acknowledgement

The author would like to thank Dr R.A. French and Dr M. Bennigsen Broxup for their assistance, although views and errors remain hers.

## REFERENCES

Akiner, S. (ed.) (1984) *Muslims of the Soviet Union*, London: Routledge & Kegan Paul.

Bennigsen, A. (1980) 'Soviet Muslims and the world of Islam', *Problems of Communism* March–April: 38–51.

Carley, P.M. (1989) 'The price of the plan: perceptions of cotton and health in Uzbekistan and Turkmenistan', *Central Asian Survey* 8(4): 1–38.

Dragadze, T. (1981) 'The sexual division of domestic space', in Ardener, S. (ed.) *Women and Space: Ground Rules and Social Maps*, London: Croom Helm.

*Ekonomicheskaya Geografia SSSR* (1973) vol. II, by Y.G. Sayshkin, I.V. Nikolski and V.P. Korovitsyn, Moscow.

Lubin, N. (1984) *Labour and Nationality in Soviet Central Asia: An Uneasy Compromise*, London: Macmillan.

Mckinder, H.J. (1904) 'The geographical pivot of history', *Geographical Journal* 23: 421–44.

Nasar, R. (1989) 'Reflections on the Aral Sea tragedy in the national literature of Turkistan', *Central Asian Survey* 8(1): 49–68.

Oreshkin, D.B. (1991) 'Ethnic dimensions of the Aral Sea crisis', paper read at a conference on *Economic Development, Ethnicity and Nationalism in the Soviet Union*, May, Cortona, Italy.

Radvanyi, L. (1990) *L'URSS: les régions et nations*, Paris: Masson.

Rywkin, M. (1982) *Moscow's Muslim Challenge: Soviet Central Asia*, New York: M.E. Sharpe.

Smith, G. (ed.) (1990) *Nationalities in the Soviet Union*, Harlow: Longman.

Symons, L. (ed.) (1990) *The Soviet Union: A Systematic Geography*, London: Hodder & Stoughton.

*The Times* (1991) 'Lake Baikal pollution threatens world systems', *The Times* 8 July: 20.

# 9

# CONCLUSIONS

## Two-speed Asia: dramatic change and stagnation

### Graham Chapman

At the beginning of this volume we highlighted several major themes concerning the coming decades in Asia, seen from the perspective of the 1960s. These were: first, food security; second, the rate of growth of the industrial economies and the relationship of this to rates of urbanization; third, the nature and style of institutional and cultural change; and fourth, linked to all the above points, the degree and manner of change in the inclusion of these states in the world economic system. As we noted, population growth rates also fascinated, or appalled, observers, but were on the whole taken to be a 'given' for many countries, with the possible exception of China. We also noted the very separate nature of Soviet Asia.

Population growth rates have for the most part passed their peaks, but they still remain extremely high by any historical comparison, though not for the most part as high as in Africa. The reasons for the slight declines observable are not easy to discover: certainly direct family planning programmes can be attributed with only a small part of the effects observed. What seems more likely in many of these societies is that cultural changes associated with urbanization and education have changed the attitudes of some couples towards the desirability and necessity of having many children. The better provision of health care in urban areas may also have contributed. These trends should obviously increase as the levels of urbanization increase, but the decline to a zero rate of population growth is still many decades off, because of the age structure of the populations, which is still overwhelmingly youthful. In Central Asia the population growth rates of the indigenous peoples remains extremely high, and throughout the period the rate was even encouraged by the authorities, who continued to award heroine mother of the Soviet Union medals to women with ten or more children. Attitudes may be on the threshold of change, but the youth of the population guarantees continuing high rates of increase for decades, even with diminishing numbers of children per family. Japan stands as a stark contrast – where the problems increasingly are of a lack of youth. Whimsically one wonders whether the automation of the car-assembly lines will now have a growing place in residential homes for the Japanese elderly.

Urbanization has not come without its problems. The inadequate provision of housing is perhaps inevitable given the rate at which cities have grown, as too the lack of proper services and sanitation might seem in some sense inevitable. But it is quite clear that policies towards informal settlements have been very contrary: sometimes they have been bulldozed away, sometimes they have been the recipients of public improvement programmes. The problems of the urban environment have not been confined to the poor only. In both lower income and especially in higher income countries, pollution by industry, local transport and power generation, both in large plants and by domestic fires, has been on a massive scale. If Japan and Taiwan are anything to go by, it does seem that some of the worst effects are ameliorated by action following public concern and legislation – which tends to support the otherwise somewhat shaky hypothesis that only wealth – which creates many of the problems – can find the resources to solve them. But it does beg the question why it is necessary for people to have to suffer the interim worst stages of pollution, and why the results of neglect should be bequeathed to the next generation. Certainly even the wealthiest societies still have many problems to deal with, of contemporary making as well as of past legacy.

The most heartening change throughout Asia has been that for most nations the threat of famine is for the time being diminished. Agricultural output has nearly everywhere increased, even if productivity levels are often woefully low. The Green Revolution for many states has been a success at least in part. The worst effects of increased social inequality have been passed, and it seems as if the technology and the capacity to use it in many states now helps the smaller farmers as much as the larger ones. In some areas there have been great improvements in land policy – either in dismantling the worst cases of state control in China or Vietnam, or by some redistribution from landlord to peasant in parts of India. But the position with regard to land-reform is still patchy, and the nature of desirable goals still unclear in many cases. Land redistribution in Bangladesh for example would hardly make any impact on the viability of many small farms – even though the current land-tenure structure in many areas of the country is still a constraint on development. The point that has often been made, that the Green Revolution has provided a breathing space for these societies, seems to have validity. But there is as yet no guarantee that agricultural productivity will continue to match or outpace population growth rates in the next decades. One of the reasons why it may be difficult for this to happen in the future is that recent increases in productivity have been so closely tied not just to agro-chemicals, but to increased provision of irrigation. The scope for large-scale schemes, and even their desirability, is rapidly diminishing, and small-scale schemes, usually involving ground water and pumping, are also becoming increasingly costly in energetic terms, and are sometimes finding that aquifers are being mined beyond sustainable limits. All of this is occurring at a time when urban areas and industry are making increasing demands on water too, and when future demands are set to escalate rapidly. The situation in China could clearly become very critical indeed, and in many parts of South Asia is already critical.

Industrial production has in all countries increased significantly, although in some cases less than hoped for, yet in other cases dramatically. Although there seems to be a highly varied number of correlates to explain why, one that seems *not* to feature prominently (except in the case of former Soviet Asia) is natural resources, one of the classic determinants in past geographies. The most successful economy of Asia, if not of the world, over the last twenty-five years has been that of Japan, which has no natural resources worth speaking of, but it does have the most valuable resource of all – possibly the world's best educated workforce. That workforce has been trained as a result of decisions taken to invest in education both by the government and by individual parents. It speaks of a cultural attitude not only about education but also about the opportunities forgone – consumption goods forgone in favour of this type of investment. There comes a point, however, at which, given a very low population growth rate and a high enough output, it becomes possible to maintain that kind of investment in education and also to enjoy the fruits of consumption – a stage which Japan appears to have reached. The same is not yet true for South Korea, where consumption is still forgone in favour of education, but presumably that may change before long.

Another correlate that seems to suggest itself is the extent to which most governments have been able to improve by substantial margins all forms of communications – whether roads, railways, telephones or cable networks. Again, the exception seems to be the former USSR, where transport has never had the priority investment due, except in occasional and ill-thought-out major schemes, such as the BAM. Though the spectacular investments have been in such networks as the Japanese Bullet Train, the precursor to France's TGV, the most pervasive influence throughout the densely settled parts of Asia has been through the growth of rural road networks, often of appalling standards and variable maintenance, but still linking communities in the Himalayas of Nepal and Bhutan, or up-country Thailand, at a speed which was not anticipated in the 1960s. This has already had discernible impacts on the pattern of rural markets, and has enabled many small farmers to benefit from occasional sales of small surpluses. The longer-term impact in terms of changing rural isolation and tradition is not yet documented. Not all countries have yet provided similar access for rural populations to cheap and reliable telephone services; indeed this telecommunications discrepancy between developed and underdeveloped nations is much wider than the discrepancy in motor-vehicle access. Yet the adequate flow of information has time and again been proven to have a significant impact on the transfer of wealth to rural areas by providing reliable market information.

The size of countries would also seem to have something to do with 'success', although not in simple fashion. A large size was once thought to be a virtue, giving large internal markets, and the possibility of the development of major growth poles which would then trickle stimulus down to peripheral regions. But this survey has shown the extent to which the smaller rather than the larger nations have been some of the outstanding growth economies, be it Singapore, or Taiwan, or Hong Kong. What they have, in common with Japan despite its greater size, is a significant degree of social homogeneity, which says much for the degree of consensus that may emerge on policy issues. So it might seem that coherence and consensus seem significant correlates of successful development, but ethnic and linguistic homogeneity alone are not enough. One of the most homogeneous states in this area in this regard is Bangladesh –

yet that homogeneity does not equate with a coherent society, but rather with a fractious one, still to many eyes with significant elements of feudalism in its agrarian structure.

The larger states, India, Indonesia, China and the USSR, have all displayed the contrary problems of size. In all, though to differing degrees, central governments have been frightened by regional autonomy, and have probably dampened the speed of overall development by imposing from the centre inappropriate policies on local areas. In the case of India at least, there has been a gradual shift in power, both by deliberate devolution to state-level planners, and also by erosion of central authority. The fact that India has at least two speeds of development predates Independence, but it is also now both part cause and part result of this devolutionary change. One might therefore ask, would it matter if the larger states split into smaller ones, each able to pursue its own policies in its own interests? There is of course no simple answer to this question. A large state does not have to connote strong central authority – it can connote the reverse, with federating units devolving upwards to the centre only those functions which are best served by the centre. In a free-market capitalist economy, chief amongst those might be a single currency and co-ordinated defence and foreign policy. This would mean changing the states such as India and China into voluntarist common markets, and acknowledging the right to secession of component parts. Such a state might even prove attractive to Pakistan and Bangladesh. It is somewhat analogous to the early stages of the 'Old (British) Commonwealth', a model frequently cited by the founding fathers of the new Commonwealth of Independent States that is emerging from the collapse of the USSR. If on the other hand states simply opt for total independence and financial autonomy, as Bangladesh, once part of Greater India, has done through the twists of history, there is every chance that the result will be an initial abortive attempt at import substitution and domestic protection, followed by a growing dependence on the world metropole. Much is made these days of the importance of south–south co-operation, of the significance of emerging organizations like ASEAN, but in fact little of the potential has been achieved so far. SAARC is an example of a very slowly moving regional accommodation. Therefore for the moment the costs of giving up the common markets that such countries as India represent should not be ignored.

A similar argument was advanced against the break-up of Soviet

Asia, but now that these states have become independent of Moscow, it is not clear that they will easily be able to relate to the wider world economy, since major deficiencies in transport links cannot be rectified quickly, and may not be viable investments anyway. External investors will not be interested in the development of the region as the Soviets were, no matter the inefficiencies of their system, rather they will be interested in investment for exploitation of resources only. The location, the low density and low income of the population mean that the region has no value for investors in manufacturing industry.

The penetration of the metropole into other areas of developing Asia is occurring despite pretensions at autarky, but it involves more and more the penetration not of the Europeans and the Americans, but of the Japanese who have come increasingly to leave their fingerprints on all the other economies covered in the volume. Other countries might decry the new imperialism which Japan could not achieve in war but has achieved in peace, but they need both the capital that Japan, the greatest aid donor, can provide, and also the technology, which so often is adapted better and faster to local conditions by the Japanese than by any other nation. The Japanese too are anxious for resources, be it Indonesia's tropical timber, or India's iron ore. The scale of these trade relationships and dependencies will pose increasingly difficult questions about the limits of policy-making for many of the other states of Asia.

The issue of the definition of an 'appropriate state' is also raised in a number of other ways. Large-scale refugee movements in Asia seem to suggest that many states have difficulty in incorporating all members within their polity. The influx of refugees that occurred into Pakistan from India at Independence has still not been resolved – the split between the incomers and the indigenous population is a major destabilizing feature of local politics. The same is true in Taiwan, where the Nationalists from the mainland have dominated the indigenous Taiwanese. More recently 'economic' refugees have flooded out of Bangladesh into Assam, precipitating conflict there which culminated in terrible massacres in 1983, and out of Vietnam by boat to Hong Kong and elsewhere. War refugees have fled from Cambodia/Kampuchea into camps in Thailand, and a similar but smaller scale of movements occurs on the Bangladesh–Burma border. All these cases suggest that the relationship between citizen and nation state has not been adequately worked out, and that it may be the nature of the state, currently still grounded in conceptions derived from nineteenth-century Europe, not the citizen, that has to change.

The circumstances for precipitating change have been altered by the end of the Cold War. This does not mean that a new era of objectivity as Myrdal wanted will be ushered in – a new set of ideological blinkers can come into play very rapidly. Already we hear that aid may come not just with the conditions of reformed economic policy attached, we are also beginning to hear of conditions of good governance attached – by which is meant adherence to Western democracy. Prima facie this is no bad thing – but how will the outside powers react if good governance means a democratic system that will work against existing state definition – independence for a landlocked Punjab, for landlocked Kashmir, for Tibet, for Inner Mongolia, for Tartarstan and Yakutia in Russia, for parts of the Philippines, for East Timor, for the Shan states of Burma? And the list could be much longer. As we write, this same issue has come to a head outside Asia over the republics of Yugoslavia, and it is possible that new precedents will soon be quoted by proponents of the Asian cases that we have mentioned. Though it might seem to avoid the question, the answer has to be in developments of many differing political structures, more or less representative at higher or lower levels, according to different cultures, histories, and economies.

There seems little doubt that whatever structures emerge, nearly all will acknowledge the role of membership of the international trading community. Whether or not aspiration to higher material standards of living is a 'good thing', for the moment at least it seems to be the major aspiration of most peoples in most states, and most acknowledge that this cannot be achieved without transfer of technology and without resort to capital borrowing from capital surplus countries. Such a pattern is not new, but is at least as old as the industrial revolution. What is new is the realization of what a fully industrialized Asia means in terms of resource demands and in terms of the output of pollution. The giants, India and China, have hardly begun to use fossil fuels to the extent that they will have to, to meet predicted energy needs, and the output of acidic pollutants and greenhouse gases is bound to rise substantially. Both are inefficient users of energy at present, China spectacularly so, hence there is an implication that with the right investment and technology transfer, much could be achieved while minimizing the increase in pollutant output. But this is not an issue which is being researched adequately at the moment either from the political or economic viewpoints. We do not know China's future political structure, and perhaps we cannot wait until it gets an EC or UK badge of good governance before helping in the process. There is also an unfortunate irony that it is Asian states, Bangladesh in particular, though much of coastal China too, which might suffer most from the rising sea-levels associated with global warming. (Although the effects of this are not easily computed, and are indeed subject to intense and unresolved debate.) We do not know either of the true extent of deforestation in Asia, despite the availability of satellite imagery. We have inadequate past records of an equivalent nature, and difficulty in establishing the 'ground truth' of much of the data. But it does seem likely that much of Asia's species diversity is at risk, and that much of its irrigated farm-land has suffered from increasing salinization.

Ecological catastrophe is not the prerogative of the densely settled states. Environmental damage inflicted in the Aral Sea region or in Lake Baikal is fundamentally a product of a political system that combined horrendously two particular traits: first, the ideological belief that nature can and will be remodelled to human demands, and second, prohibition of public debate, dissent and bottom-up feedback.

Thus the transformation of Asia in many parts over the last twenty-five years has been dramatic – as Myrdal implied that he hoped it would be. But in Soviet Asia the last twenty-five years are part of the wasted paralysis of the era from Khrushchev to Brezhnev, when little was achieved except increased damage. Only now is the possibility of real and beneficial change dawning, but such change is not guaranteed. For yet other parts, particularly in Low Speed India, parts of Pakistan and Bangladesh, and in parts of China and Indo-China, there are literally hundreds of millions of people still in appalling poverty, and for whom improvement must seem generations away. In these areas and in the many other areas of the Island States of Indonesia and the Philippines, the escape from poverty is awaiting changes in the social system, and then the political system it supports. It does not help if the political system is ostensibly democratic – if a power structure still buried in the social system of the past can continue to dominate the 'new' system. For the wealthy of Asia the first issue of importance is the security of future resource supplies, many of which come from poor regions. Thus they too should now recognize the importance of sustainable development for Asia as a whole, and indeed as the West should recognize, the importance of this for the planet.

# Index